Mechanics and Adaptronics

Ziel der Buchreihe ist die Publikation neuer Ansätze in der Mechanik und Adaptronik. Diese Ansätze umfassen zum einen experimentelle Themen als auch solche, die numerischen Charakters sind und erstrecken sich über diverse Größenskalen. Experimentelle, zukunftsweisende Themen beinhalten Untersuchungen auf Mikro- bis hin zur Makroebene und berücksichtigen unterschiedlichste Felder, von der Mechanik über Temperatur bis hin zu unterschiedlichen chemischen Feldern. Auch auf Seiten der numerischen Simulation werden auf Grundlage der experimentellen Untersuchungen neue multiskalige Mehrfeldmodelle entwickelt, kalibriert und validiert. Eine Besonderheit stellt hier die Thematik der Adaptronik dar, die als Ziel die Integration neuer Funktionen in tragende Bauteile durch die Kombination konventioneller Konstruktionswerkstoffe mit aktiven Werkstoffsystemen verfolgt. So werden Funktionswerkstoffe betrachtet, die Energiewandler sind und die die klassischen lasttragenden und formgebenden Funktionen um sensorische und aktorische Funktionen erweitern. Aktorische und sensorische Eigenschaften ermöglichen in Kombination mit adaptiven, schnellen Reglern eine optimale Selbstanpassung an die jeweilige Einsatzumgebung. Durch die Zusammenführung von numerischem und experimentellem Know-how auf mechanischer als auch adaptronischer Seite ist es möglich, Erkenntnisse aus der Natur im Sinne des bionischen Gedankens in die Technik zu übertragen und so neue wissenschaftliche Richtungen aufzuschlagen.

Johannes Wiedemann

Relevance of Manufacturing-Induced Residual Stress States in Fiber Metal Laminates for Structural Health Monitoring

Springer

Johannes Wiedemann
Institut für Mechanik und Adaptronik
Technische Universität Braunschweig
Braunschweig, Germany

ISSN 2731-6211 ISSN 2731-622X (electronic)
Mechanics and Adaptronics
ISBN 978-3-032-09924-2 ISBN 978-3-032-09925-9 (eBook)
https://doi.org/10.1007/978-3-032-09925-9

Diese Veröffentlichung wurde aus Mitteln des Publikationsfonds NiedersachsenOPEN, gefördert aus zukunft.niedersachsen, unterstützt.
The author further acknowledges support by the Open Access Publication Funds of Technische Universität Braunschweig.
This work was additionally supported by Helmut-Schmidt-Universität / Universität der Bundeswehr Hamburg and the German Research Foundation (Deutsche Forschungsgemeinschaft (DFG)) within the Research Unit 3022 "Ultrasonic Monitoring of Fibre Metal Laminates Using Integrated Sensors".

This Springer imprint is published by the registered company Springer Nature Switzerland AG
The registered company address is: Gewerbestrasse 11, 6330 Cham, Switzerland

If disposing of this product, please recycle the paper.

Relevance of Manufacturing-Induced Residual Stress States in Fiber Metal Laminates for Structural Health Monitoring

Von der Fakultät für Maschinenbau
der Technischen Universität Braunschweig
zur Erlangung der Würde

eines Doktor-Ingenieurs (Dr.-Ing.)

genehmigte Dissertation

von: Johannes Wiedemann
geboren in: Mammendorf

eingereicht am: 05.12.2024
mündliche Prüfung am: 28.02.2025

Vorsitz: Prof. Dr. rer. nat. Andreas Dietzel
Gutachter: Prof. Dr.-Ing. Christian Hühne
Gutachterin: PD Dr.-Ing. habil. Natalie Rauter

2026

Abstract

Structural health monitoring (SHM) is a promising technology for monitoring the structural integrity of fiber metal laminates (FMLs) during operation. SHM systems using guided ultrasonic waves (GUWs) enable the detection of damage located inside a laminate and are, therefore, particularly suitable for FMLs due to their complex damage characteristics. This dissertation explores how manufacturing-induced residual stress states in FMLs affect such an SHM system.

Experimental methods for quantifying residual stresses in FMLs are developed and evaluated. This work shows that process strains can be measured throughout manufacturing with sensors integrated into the laminate. The strain measurements verify different modified cure cycles to change residual stress states in FMLs consisting of carbon-fiber-reinforced polymer (CFRP) and steel. To avoid complex sensor technology in industrial manufacturing processes, a robust method is additionally presented that derives a process-specific parameter, the stress-free temperature, from asymmetrical reference laminates. The stress-free temperature can quantify residual stress states of other laminates with a symmetrical layer structure produced in the same manufacturing process.

The quantified residual stress states in various CFRP-steel laminates allow the experimental investigation of the influence of stress on the propagation velocity of fundamental GUWs. With laser-scanning-vibrometry, dispersion diagrams for various FMLs over a wide frequency range are experimentally determined. The results show no significant influence of the residual stress states on the phase velocity of the fundamental symmetric and asymmetric GUW modes for the investigated FMLs. In contrast, comparative measurements under external tensile prestress show a considerable increase in phase velocity with increasing prestress for both fundamental GUW modes over a wide frequency range.

Finally, fatigue tests are performed on CFRP-steel laminates to investigate the influence of different residual stress states on the residual life prediction capabilities of an SHM system. The results show that reducing residual stresses through modified cure cycles delays crack initiation, reduces crack propagation velocity, and, consequently, extends the fatigue life of the investigated FMLs. Therefore, the results indicate that the fatigue strength of CFRP-steel laminates under cyclic loading can be improved using modified cure cycles.

The findings of this dissertation lead to two main consequences in the context of SHM systems for FMLs. First, since wave propagation velocity is important for damage localization using GUWs, the influence of external loads must be considered in contrast to the influence of manufacturing-induced residual stress states. Second, for remaining service life predictions after damage detection, the knowledge of the residual stress state in FMLs helps improve the prediction quality. The results contribute to advancing the technological maturity of SHM systems for FMLs by demonstrating fundamental relationships between residual stress states and relevant variables of an SHM system and by quantifying the effects.

Kurzfassung

Die integrierte Strukturüberwachung (engl.: Structural Health Monitoring, SHM) stellt eine vielversprechende Technologie dar, um die strukturelle Integrität von Faser-Metall-Laminaten (FMLs) während des Betriebs zu überwachen. SHM-Systeme, die geführte Ultraschallwellen (GUWs) nutzen, ermöglichen die Detektion von Schäden, die sich im Inneren eines Laminats befinden und sind daher aufgrund der komplexen Schadenscharakteristik von FMLs für diese besonders geeignet. Diese Dissertation untersucht, wie sich die inhärenten fertigungsbedingten Eigenspannungszustände eines FMLs auf ein solches SHM-System auswirken.

Dafür werden experimentelle Methoden zur Quantifizierung von Eigenspannungen in FMLs entwickelt und evaluiert. Die Arbeit zeigt, dass mit Hilfe von im Laminat integrierter Sensorik die entstehenden Prozessdehnungen über den gesamten Fertigungsprozess hinweg gemessen werden können. Mit der entwickelten Methodik werden verschiedene modifizierte Aushärtezyklen zur gezielten Einstellung von Eigenspannungszuständen in FMLs, bestehend aus kohlenstofffaserverstärktem Kunststoff (CFK) und Stahl, überprüft und validiert. Um in industriellen Fertigungsprozessen auf aufwendige Sensorik verzichten zu können, wird darüber hinaus eine robuste Methode vorgestellt, die aus asymmetrischen Referenzlaminaten einen prozessspezifischen Parameter, die spannungsfreie Temperatur, ableitet. Die spannungsfreie Temperatur ermöglicht die Quantifizierung des Eigenspannungszustands auch für andere Laminate mit symmetrischem Lagenaufbau, die im selben Fertigungsprozess hergestellt werden.

Die quantifizierten Eigenspannungszustände in verschiedenen CFK-Stahl-Laminaten erlauben es, deren Einfluss auf die Ausbreitungsgeschwindigkeit von fundamentalen GUWs experimentell zu untersuchen. Mittels Laser-Scanning-Vibrometrie werden für die verschiedenen FMLs Dispersionsdiagramme über einen weiten Frequenzbereich experimentell ermittelt. Die Ergebnisse zeigen, dass in den untersuchten FMLs kein signifikanter Einfluss der Eigenspannungszustände auf die Phasengeschwindigkeit der fundamentalen symmetrischen und asymmetrischen GUW-Moden nachweisbar ist. Im Gegensatz dazu zeigen Vergleichsmessungen unter äußerer Zugvorspannung eine erhebliche Zunahme der Phasengeschwindigkeit mit zunehmender Vorspannung für beide fundamentale GUW-Moden über breite Frequenzbereiche.

Abschließend werden Ermüdungsversuche an verschiedenen CFK-Stahl-Laminaten durchgeführt, um den Einfluss verschiedener Eigenspannungszustände auf die Restlebensdauervorhersage eines SHM-Systems zu untersuchen. Die Ergebnisse zeigen, dass eine Reduktion der Eigenspannungen durch modifizierte Aushärtezyklen zu einer Verzögerung der Rissinitiierung und Verringerung der Rissausbreitungsgeschwindigkeit und damit zu einer Erhöhung der Ermüdungslebensdauer führt. Die Verwendung von modifizierten Zyklen während der Fertigung kann daher die Ermüdungsfestigkeit eines FMLs unter zyklischer Belastung im Betrieb verbessern.

Im Kontext von SHM-Systemen für FMLs ergeben sich aus dieser Dissertation zwei Konsequenzen. Erstens, da für die Schadenslokalisation mit Hilfe von GUWs die Wellenausbreitungsgeschwindigkeit von großer Bedeutung ist, muss der Einfluss äußerer Lasten, im Gegensatz zum Einfluss von fertigungsinduzierten Eigenspannungszuständen, berücksichtigt werden. Zweitens, für Restlebensdauerprognosen nach Schadensdetektion in einem FML ist die Kenntnis des Eigenspannungszustands hilfreich, um die Prognosegüte zu verbessern. Die gewonnenen Ergebnisse tragen dazu bei, die technologische Reife von SHM-Systemen für FMLs voranzutreiben, indem grundlegende Zusammenhänge zwischen Eigenspannungszuständen und relevanten Größen eines SHM-Systems aufgezeigt und die resultierenden Effekte quantifiziert werden.

Acknowledgements

This dissertation was written during my time at the Institute of Mechanics and Adaptronics of the Technical University Braunschweig and was supported by the German Research Foundation (Deutsche Forschungsgemeinschaft – DFG) as part of the Research Unit FOR3022. I want to express my gratitude to the DFG for funding my research.

I would especially like to thank the examination committee – particularly my doctoral advisor Christian Hühne – for his unwavering support, the freedom to pursue my ideas, and many enlightening technical discussions. My thanks also go to Natalie Rauter for serving as the second reviewer, for her thorough evaluation of my work, and for the constructive academic exchange throughout the research unit. I further extend my gratitude to Andreas Dietzel for chairing the examination committee and for his stimulating contributions during our group meetings.

I am furthermore grateful to all the research group members, whose contributions were integral to the success of this work. The regular meetings and lively discussions in a large group constantly motivated and energized me throughout my doctoral studies. I wish to especially highlight Michael Sinapius and Wolfgang Weber, whose leadership and encouragement inspired me to continuously push the boundaries.

Beyond the research group, I am grateful to Selim Mrzljak from the Technical University of Dortmund for his keen interest in my work and the productive collaboration addressing the fatigue of FML. Both the scholarly and personal exchanges were truly enriching. I also thank my colleagues Bernhard Plank and Lukas Gahleitner from the University of Applied Sciences Upper Austria for the straightforward discussions and constructive collaboration in the field of nondestructive testing.

On the side of the German Aerospace Center (Deutsches Zentrum für Luft- und Raumfahrt – DLR) in Braunschweig, I extend my gratitude to Erik Kappel and Robert Prussak, who sparked my enthusiasm for the challenges in manufacturing fiber-reinforced composites and fiber-metal laminates already during my Master studies and remained always available for academic discussion during my doctoral research. I am particularly thankful to Mareike Stegmaier for sharing her expertise in the manufacturing of fiber-reinforced composites and fiber metal laminates, as well as for her support in producing the many specimens examined in this work.

I also wish to thank the entire team of the Institute of Mechanics and Adaptronics for fostering an inspiring work environment and mutual motivation. Special thanks go to Thomas Roloff for the insightful discussions on guided waves and for the excellent collaboration within the research group. I am equally grateful to all the students I had the pleasure of supervising and mentoring during my time at the institute – their enthusiasm made the experience exceptionally rewarding.

Finally, I express my deepest appreciation to my family, friends, and my partner, Dana Hellkamp, whose unwavering support, both inside and beyond academia, made this doctoral journey possible.

Contents

Acronyms xvii

1 Motivation 1
References 4

2 Material Characteristics and Structural Health Monitoring of Fiber Metal Laminates 7
2.1 Fiber Metal Laminates 7
2.2 Residual Stresses in Fiber Metal Laminates 10
2.2.1 Mechanical Property Evolution of an Epoxy Resin During Cure 12
2.2.2 Stress-Free Temperature 14
2.2.3 Analytic Description of Effective Ply Stresses 15
2.2.4 Modification of Residual Stress States 17
2.3 Guided Ultrasonic Wave-Based Structural Health Monitoring of Fiber Metal Laminates 19
2.3.1 Dependencies Between Residual Stress and Structural Health Monitoring 21
2.3.2 Stress Influence on Guided Ultrasonic Wave Propagation ... 22
2.3.3 Residual Stress Influence on FML Fatigue 23
2.4 Research Hypotheses and Thesis Outline 24
References 27

3 Materials and Manufacturing 35
3.1 Elastic Material Properties 35
3.2 Thermal Material Properties 37
3.3 Steel Sheet Preparation and Layup 39
3.4 Cure Cycles 42
3.5 Resin Cure Kinetics 43
3.6 Preliminary Investigations - Laminate Strength and Cure Cycle Influence 45

References 47

4 Residual Stress Quantification Using Strain Gauges 51
4.1 Motivation 51
4.2 Strain Gauge Technique 53
4.2.1 Evaluation of Measured Strains During Cure 55
4.2.2 Determination of Bonding Temperatures 60
4.2.3 Residual Stress Determination 61
4.3 Validation of Strain Gauge Measurements With Fiber Bragg Grating Sensors 63
4.4 Estimation of the Validity of Single Point Measurements 66
4.4.1 Experimental Temperature Distribution Across the Thickness in FML 67
4.4.2 Numerical Modeling of the Through-Thickness Temperature Distribution in FRP Laminates 68
4.5 Conclusion and Discussion 72
References 74

5 Residual Stress Quantification Using Asymmetric Laminates 79
5.1 Motivation 79
5.2 Experimental Setup and Specimen Program 82
5.3 Experimental Methods for Parameter Determination From Asymmetric Specimens 84
5.3.1 Curvature Evaluation 84
5.3.2 Stress-Free Temperature Evaluation 84
5.4 Results of the Experimental Curvature Analysis 85
5.4.1 Intrinsic Parameter Influences 86
5.4.2 Extrinsic Parameter Influences 88
5.4.3 Correlation of Curvature and Stress-Free Temperature 90
5.4.4 Validation Using the Strain Gauge Method 93
5.5 Residual Stress Levels for Relevant Laminate Configurations 94
5.6 Conclusion and Discussion 95
References 97

6 Residual Stress Influence on Guided Ultrasonic Wave Propagation 99
6.1 Motivation 99
6.2 Experimental Method to Determine Guided Ultrasonic Wave Dispersion Diagrams 102
6.3 Validation of the Experimental Setup 104
6.3.1 System Reproducibility 105
6.3.2 Specimen Geometry 106
6.3.3 Comparison to Analytical Solution 107
6.4 Comparison of Dispersion Diagrams for Different Residual Stress States 108
6.5 Comparison of Dispersion Diagrams for Different Prestress States .. 110

6.6 Conclusion and Consequences for SHM 113
References 116

7 Residual Stress Influence on Fatigue Life 119
7.1 Motivation 119
7.2 Experimental Setup and Evaluation 121
7.3 Experimental Fatigue Damage Evolution in CFRP-Steel Laminates . 123
7.4 Residual Stress Influence on Fatigue Life Phases 126
7.4.1 Crack Initiation 128
7.4.2 Crack Growth 129
7.5 Laminate Parameters Influencing Fatigue Damage Evolution 131
7.6 Conclusion and Consequences for SHM 132
References 134

8 Conclusion and Outlook 137
8.1 Main Contributions of the Dissertation 137
8.2 Evaluation of the Research Hypotheses 140
8.3 Open Questions and Future Work 141
References 143

Acronyms

AITM	Airbus Industry Test Method
ARALL	Aramid-fiber-reinforced aluminum laminate
ASTM	American Society for Testing and Materials
BVID	Barely visible impact damage
CARALL	Carbon-fiber-reinforced aluminum laminate
CAT	Constant amplitude test
CBM	Condition-based maintenance
CFRP	Carbon-fiber-reinforced polymer
CLT	Classical laminate theory
CT	Computer tomography
CTE	Coefficient of thermal expansion
DIC	Digital image correlation
DMA	Dynamic mechanical analysis
DSC	Differential scanning calorimetry
FAA	Federal Aviation Administration
FBG	Fiber Bragg grating
FEA	Finite element analysis
FML	Fiber metal laminate
FOS	Fiber optic sensor
FRP	Fiber-reinforced polymer
FTF	Facesheet thickness fraction
GFRP	Glass-fiber-reinforced polymer
GLARE	Glass-fiber-reinforced aluminum laminate
GUW	Guided ultrasonic wave
HTC	Heat transfer coefficient
HTCL	Hybrid titanium composite laminate
LSV	Laser scanning vibrometer
MRCC	Manufacturer recommended cure cycle
MOD	Modified cure cycle
MRO	Maintenance repair and overhaul
MVF	Metal volume fraction

NoI	Number of interfaces
NT	Nodal temperature
OHT	Open hole tension
PTFE	Polytetrafluoroethylene
RSD	Relative standard deviation
SEM	Scanning electron microscopy
SG	Strain gauge
SHM	Structural health monitoring
TC	Thermocouple
TiGr	Titanium graphite
ToF	Time-of-flight
T_{sf}	Stress-free temperature
ULE	Ultra-low expansion
XRD	X-ray diffraction

Chapter 1
Motivation

Fiber metal laminates (FMLs) were initially developed to increase the fatigue and damage tolerance of monolithic metallic structures in the aerospace industry. Adding layers of glass-fiber-reinforced polymer (GFRP) between thin aluminum sheets (0.2 mm to 0.5 mm) lead to a new structural material with significantly improved fatigue characteristics, and finally to the industrial application of GLARE (glass-fiber-reinforced aluminum laminate) in the Airbus A380. Compared to monolithic aluminum, GLARE shows a significantly higher tolerance to fatigue and damage due to the layered architecture. GLARE generally contains a high metal volume fraction (MVF) and is therefore classified as a metallic material for certification. [1,22]

While GLARE is considered a metallic material, it was demonstrated that the mechanical performance of fiber-reinforced polymers (FRPs) can be improved by substituting single plies of monolithic FRPs with metallic layers. Substituting the fiber layers oriented transverse to the primary load path with metal sheets results in composites with higher joint robustness and efficiency due to their increased load-bearing capacity [5, 10]. The advantage here is the possibility of local hybridization which only uses the metal locally in the highly loaded area, such as the bolted connection, leading to low metal volume fractions in the overall composite structure [3, 16]. These advancements were mainly proven for carbon-fiber-reinforced polymers (CFRPs) in combination with titanium or stainless steel.

Merging the two single materials in a laminated structure, as shown in Fig. 1.1, leads to a material that can compensate for the disadvantages of the single constituents, by combining the high specific strength and stiffness of FRPs with the

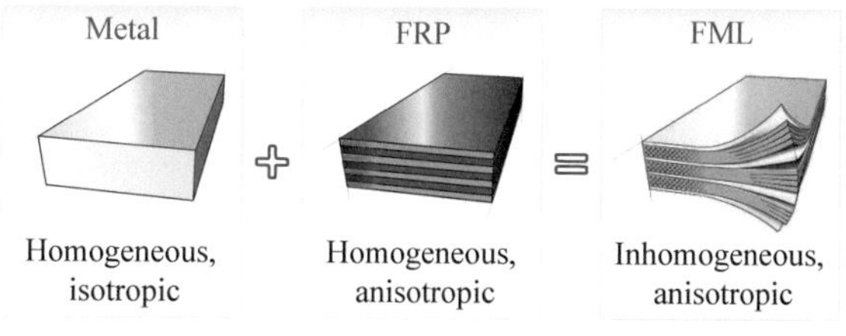

Fig. 1.1 The layered combination of metals with FRP produces a material with superior mechanical properties but high anisotropy and inhomogeneity.

J. Wiedemann, *Relevance of Manufacturing-Induced Residual Stress States in Fiber Metal Laminates for Structural Health Monitoring*, Mechanics and Adaptronics, https://doi.org/10.1007/978-3-032-09925-9_1

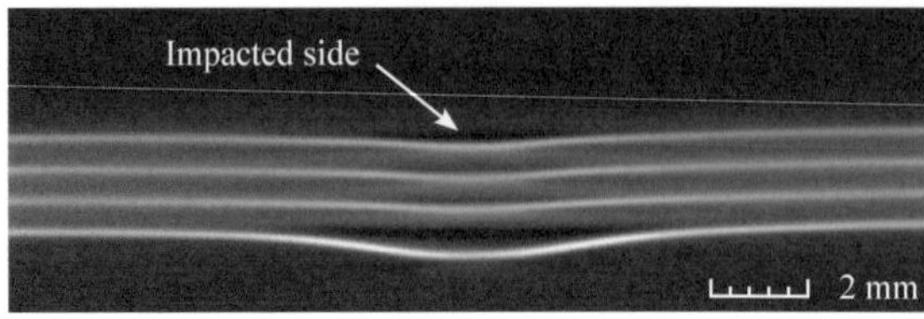

Fig. 1.2 Computer tomographic image of a CFRP-steel laminate damaged with a low-velocity impact (impact energy: 5 J, impactor diameter: 15.9 mm).

ductility and damage tolerance of metals. Consequently, many combinations of FRPs with metals have been investigated in the literature, showing that these combinations can outperform monolithic materials, especially in highly loaded areas and areas prone to impact damage. [20, 22]

While FMLs feature distinct mechanical advantages, their inspection in the event of damage is challenging. Due to the layered structure of FMLs (Fig. 1.1), they exhibit complex damage characteristics with damage patterns that are often difficult to identify through visual inspection techniques from the outside, such as delaminations between different layers after impact damage. Figure 1.2 shows the computer tomography (CT) cross-sectional image[1] of an FML damaged with a low-velocity impact. The figure reveals that significantly more damage occurs away from the impacted side and inside the laminate. This underscores the need for non-destructive testing methods to identify such barely visible impact damage (BVID) and to assure the structural integrity of FMLs.

A promising method to monitor the structural integrity and state of damage is the implementation of structural health monitoring (SHM) systems [8]. Besides identifying damage, an SHM system should be able to locate the damage, assess its impact on the structural integrity, and predict the remaining service life of the component [19]. For aerospace applications involving composite structures with large in-plane dimensions and limited thickness, guided ultrasonic wave (GUW) based SHM systems effectively address these four tasks [6, 11].

However, in the case of FMLs, additional complexity is added as they show a highly anisotropic and inhomogeneous structure due to the significant differences in stiffness between metal and FRP (cf. Fig. 1.1). This stiffness difference produces large inhomogeneities in the thickness direction with significant jumps in impedance, leading to complex in-plane displacements of propagating GUWs [17]. Nevertheless, the literature indicates that fundamental assumptions for GUW propagation in layered media are also valid for FMLs [2, 15, 18] and damage detection using SHM systems based on GUWs is possible [12, 13].

SHM systems using GUW must solve an inverse problem and, therefore, require a unique correlation between damage and a change in the wave signal [7]. However, different environmental and operational parameters in real applications like temperature, moisture, or static and dynamic loads can influence the damage-sensitive features of propagating GUWs [21]. In this context, the thermally induced interlaminar residual stress state in FMLs must also be considered an influencing factor. These

[1] The CT scans on FMLs were performed in cooperation with Bernhard Plank from the Research Center Wels of the University of Applied Sciences Upper Austria.

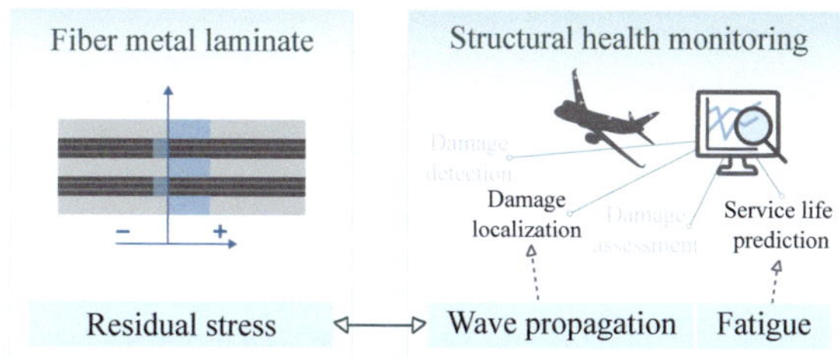

Fig. 1.3 This dissertation focuses on the influence of residual stresses in FMLs on wave propagation and fatigue, which directly affects two tasks of an SHM system: damage localization and service life prediction.

stresses are generated during the manufacturing process due to the process temperature, the significant differences in the coefficients of thermal expansion (CTEs), and the stiffness difference between metal and FRP. These residual stresses in FMLs usually consist of tensile stresses in the metal layers and compressive stresses in the FRP layers.

The primary research question of this thesis is whether the residual stresses introduced into an FML during manufacturing significantly influence an SHM system based on GUW. To address this, CFRP-steel laminates with varying residual stress states are manufactured using modified cure cycles. The residual stress states resulting from these manufacturing processes are quantified using two different methods. This analysis serves as the foundation for the subsequent investigations, which propose that the residual stress state may impact an SHM system thorugh to two main effects. First, internal stresses or strains within the material affect the velocity of propagating GUWs, as demonstrated for homogeneous materials [4]. This phenomenon is known as the acoustoelastic effect and can be explained by non-linear elastic theory [14]. Second, the interlaminar residual stress state in FMLs induces premature material fatigue [1]. The primary cause is the tensile prestress in the metal sheets, which reduces the maximum tolerable load and thus decreases the fatigue strength [9]. These two effects have not been comprehensively investigated in the context of SHM systems designed for FML structures. Based on these two effects, damage localization, and service life prediction are expected to be the two tasks of an SHM system most influenced by residual stresses (Fig. 1.3). Therefore, this dissertation aims to determine whether these effects are significant enough to be considered in solving the tasks of an SHM system.

This thesis is structured as follows. Chap. 2 outlines the current state of research related to the motivation of this work and formulates hypotheses based on the findings. While Chap. 3 introduces the relevant materials and manufacturing processes used, the Chaps. 4-7 of this dissertation are divided into two parts, as shown in Fig. 1.3. The first part (Chaps. 4 and 5) discusses the comprehensive quantification of residual stress states in FMLs using different experimental methods and modified cure cycles. In the second part (Chaps. 6 and 7), the influence of the quantified residual stress states in various FML configurations on wave propagation and fatigue characteristics is evaluated. Finally, Chap. 8 evaluates the research hypotheses based on the findings in these four chapters.

References

1. Alderliesten, R.: Fatigue and fracture of fibre metal laminates, *Solid mechanics and its applications*, vol. 236. Springer, Cham (2017)
2. Barth, T., Wiedemann, J., Roloff, T., Hühne, C., Sinapius, M., Rauter, N.: Investigations on guided ultrasonic wave dispersion behavior in fiber metal laminates using finite element eigenvalue analysis. PAMM **23**(1) (2023). doi:10.1002/pamm.202200149
3. Both, J.C.: Tragfähigkeit von CFK-Metall-Laminaten unter mechanischer und thermischer Belastung: Zugl.: München, Techn. Univ., Diss., 2014. Ingenieurwissenschaften. Dr. Hut, München (2014). URL `http://mediatum.ub.tum.de/node?id=1173300`
4. Chen, F., Wilcox, P.D.: The effect of load on guided wave propagation. Ultrasonics **47**(1-4), 111–122 (2007). doi:10.1016/j.ultras.2007.08.003
5. Fink, A., Camanho, P.P., Andrés, J.M., Pfeiffer, E., Obst, A.: Hybrid CFRP/titanium bolted joints: Performance assessment and application to a spacecraft payload adaptor. Composites Science and Technology **70**(2), 305–317 (2010). doi:10.1016/j.compscitech.2009.11.002
6. Giurgiutiu, V.: Structural health monitoring with piezoelectric wafer active sensors. Academic Press/Elsevier, Amsterdam (2008). URL `http://site.ebrary.com/lib/alltitles/docDetail.action?docID=10203616`
7. Giurgiutiu, V.: Structural health monitoring of aerospace composites. Academic Press is an imprint of Elsevier, London, UK (2015). URL `http://www.sciencedirect.com/science/book/9780124096059`
8. Güemes, A., Fritzen, C.P., Balageas, D.: Structural health monitoring. ISTE, London and Newport Beach, CA (2006). doi:10.1002/9780470612071
9. Homan, J.J.: Fatigue initiation in fibre metal laminates. International Journal of Fatigue **28**(4), 366–374 (2006). doi:10.1016/j.ijfatigue.2005.07.030
10. Kolesnikov, B., Wilmes, H., Herrmann, A., Pabsch, A.: Composite material with a reinforced connecting area (20.03.1999). URL `https://patents.google.com/patent/WO2000056541A1/en`
11. Lammering, R., Gabbert, U., Sinapius, M., Schuster, T., Wierach, P. (eds.): Lamb-Wave Based Structural Health Monitoring in Polymer Composites. Research Topics in Aerospace. Springer International Publishing, Cham (2017). doi:10.1007/978-3-319-49715-0
12. Maghsoodi, A., Ohadi, A., Sadighi, M., Amindavar, H.: Damage detection in multilayered fiber–metal laminates using guided-wave phased array. Journal of Mechanical Science and Technology **30**(5), 2113–2120 (2016). doi:10.1007/s12206-016-0418-9
13. Masurkar, F.: Damage detection in hybrid metal-composite plates using ultrasonic guided waves based on outliers estimate. In: Structural Health Monitoring, Materials Research Proceedings, pp. 158–165. Materials Research Forum LLC (2023). doi:10.21741/9781644902455-20
14. Ogden, R.W., Ogden, R.W.: Non-linear elastic deformations, 1. publ., unabridged and corr. republ edn. Dover books on physics. Dover Publ, Mineola, NY (1997). URL `http://www.loc.gov/catdir/description/dover031/97016162.html`
15. Pant, S.: Lamb Wave Propagation and Material Characterization of Metallic and Composite Aerospace Structures for Improved Structural Health Monitoring (SHM). Dissertation, Carleton University, Ottawa, Ontario (2014). doi:10.22215/etd/2014-10409
16. Petersen, E., Stefaniak, D., Hühne, C.: Experimental investigation of load carrying mechanisms and failure phenomena in the transition zone of locally metal reinforced joining areas. Composite Structures **182**(6), 79–90 (2017). doi:10.1016/j.compstruct.2017.09.002
17. Rauter, N., Mikhaylenko, A., Bellam-Muralidhar, N.K., Lorenz, D., Lammering, R.: Numerical simulation of guided ultrasonic waves in fiber metal laminates including model reduction methods. In: H. Waubke, P. Balazs (eds.) Tagungsband, DAGA 2021 - 47. Jahrestagung für Akustik, pp. 920–923. Deutsche Gesellschaft für Akustik e.V, Berlin (2021)
18. Rittmeier, L.: Charakterisierung des strukturdynamischen Verhaltens von Faser-Metall-Laminaten unter Anregung geführter Ultraschallwellen. Springer Berlin Heidelberg, Berlin, Heidelberg (2024). doi:10.1007/978-3-662-69051-2

19. SAE International: Guidelines for Implementation of Structural Health Monitoring on Fixed Wing Aircraft. doi:10.4271/ARP6461A
20. Sinmazçelik, T., Avcu, E., Bora, M.Ö., Çoban, O.: A review: Fibre metal laminates, background, bonding types and applied test methods. Materials & Design **32**(7), 3671–3685 (2011). doi:10.1016/j.matdes.2011.03.011
21. Sohn, H.: Effects of environmental and operational variability on structural health monitoring. Philosophical transactions. Series A, Mathematical, physical, and engineering sciences **365**(1851), 539–560 (2007). doi:10.1098/rsta.2006.1935
22. Vlot, A., Gunnink, J.W.: Fibre metal laminates: An introduction. Springer Netherlands, Dordrecht (2001). URL `https://ebookcentral.proquest.com/lib/gbv/detail.action?docID=3565888`

Chapter 2
Material Characteristics and Structural Health Monitoring of Fiber Metal Laminates

Abstract FMLs have been investigated in various combinations of FRPs and metals. Due to their layered structure, they exhibit superior damage tolerance and fatigue resistance. However, significant differences in the coefficients of thermal expansion between FRP and metal result in interlaminar residual stresses between the single layers of the FML. The literature discusses different methods to reduce these residual stresses, from which the modification of the cure cycle shows superior potential. Previous research on composite materials indicates that SHM using GUWs is a promising method for damage assessment in FMLs. Due to the dependencies of stresses and the wave propagation velocity of GUWs, as well as the influence of residual stresses on the fatigue behavior of FMLs, it is assumed that, in particular, two subtasks of an SHM system are most likely to be affected by a residual stress state in an FML. The focused literature study identifies open research questions in the interaction of FMLs with SHM that are the basis to formulate precise research hypotheses and the structure of this doctoral thesis.

2.1 Fiber Metal Laminates

FMLs aim to combine the advantages of FRP with those of metals by alternately layering the two materials in a composite laminate (Fig. 1.1). The literature highlights various motivations that have driven research to explore this type of hybrid material. The most significant motivations are illustrated in Fig. 2.1.

The most prominent motivation (1) is the improvement of the fatigue and damage characteristics of monolithic aerospace aluminum, leading to the creation of various fiber-reinforced aluminum laminates such as ARALL (aramid fibers), GLARE (glass fibers), and CARALL (carbon fibers). Due to the superior fatigue characteristics of GLARE compared to monolithic aluminum, it is extensively used in the Airbus A380 long-range passenger aircraft [93, 110].

Concurrently, there was a focus on titanium-based hybrid laminates for supersonic aircrafts (2) due to the inability of conventional aerospace metals and FRP to meet the

J. Wiedemann, *Relevance of Manufacturing-Induced Residual Stress States in Fiber Metal Laminates for Structural Health Monitoring*, Mechanics and Adaptronics, https://doi.org/10.1007/978-3-032-09925-9_2

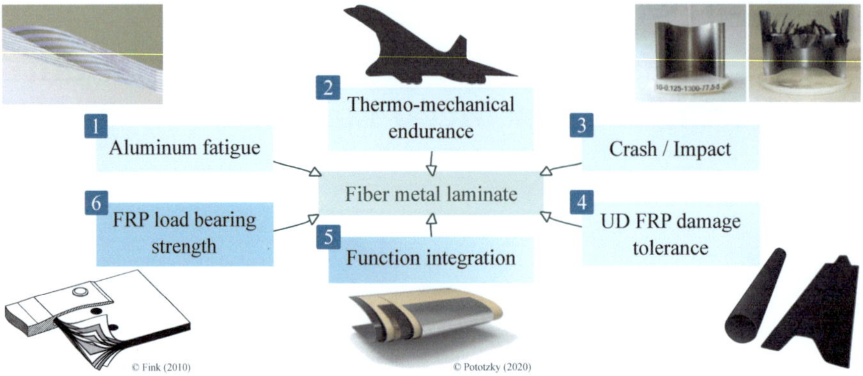

Fig. 2.1: The literature outlines various motivations for the application of FMLs. While (1) to (4) propose an FML as a structural material, in (6), the hybridization is only applied locally. Depending on the application, in (5) the hybridization might extend over the entire structure or be only applied locally. (Picture credits: [30, 91])

stringent requirements of high specific strength, stiffness, high thermo-mechanical endurance, and slow crack propagation at temperatures up to 175 °C at speeds of Mach 2.4 [127]. These laminates were commonly referred to as HTCL (hybrid titanium composite laminates) [78] or TiGr (titanium graphite) laminates [14, 71].

Many publications also investigate the superior energy absorption characteristics (3) of FMLs. For structures prone to low or high-velocity impacts, like cargo floors and leading edges, it was shown that FMLs significantly increase the damage tolerance [15, 29, 103]. In crash applications, the focus is on a high energy dissipation potential, and the combination of CFRP with steel showed a significant increase in energy absorption even with small amounts of metal in the laminate [11, 41].

Others [63, 114] showed that by incorporating low metal volume contents into primarily uni-directional oriented FRP laminates, the unique benefits and specific properties of the composite can be maximized (4). This type of FML is specifically attractive for structures dominantly loaded in a single direction. Incorporating very thin metal sheets significantly increases impact and damage tolerance due to the metal constituent and through crack stopping at the FRP-metal interfaces.

The above approaches mainly aim at the mechanical characteristics of the material combinations. At the same time, some researchers have also investigated exploiting the promising electrical and electromagnetic properties of the metal layers (5) besides their load-bearing function to create a multifunctional material [27, 28, 43, 92].

Finally, hybrid laminates were developed to address the limitations of monolithic FRP materials, such as low load-bearing capabilities (6). This is due to the lack of plastic stress redistribution known from metallic materials and the damage to the fibers after drilling. As a result, researchers explored local hybridization with titanium or steel sheets to replace FRP layers in bolted areas of a laminate. These investigations led to significant improvements in load transfer [12, 30]. The metal ply thickness is adapted to the FRP-ply thickness to allow for local ply substitutions and

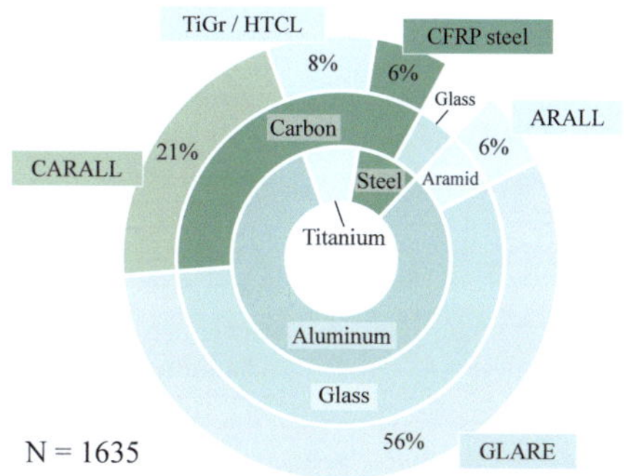

Fig. 2.2 The most common FMLs are based on combining one of three metals (aluminum, steel, and titanium) with one of three fiber materials (glass, carbon, and aramid). This is based on a systematic Scopus literature search, which identified 1635 publications dealing with FMLs made from one of these material systems.

graded hybridization with higher metal volume fractions (MVFs) around the bolt and decreasing MVFs further away from the bolt while maintaining a constant laminate thickness and component volume [64, 89]. Local steel hybridization effectively increases load-bearing strength even in GLARE [120] and GFRP laminates [35].

A Scopus literature search was conducted to reveal the most relevant material combinations in FML research. A total number of 2302 publications[1] was found. The search excludes all publications that only refer to hybrid materials since a hybrid material does not necessarily have to be a laminate and includes many more possibilities for combining fiber materials and metals. An FML is, therefore, a hybrid material, but not every hybrid material is an FML. Furthermore, from the 2302 publications, all the publications that do not specify the material systems in the title, abstract, or keywords were excluded. This is often the case in numerical studies where only a generic FML is used without specifying the material combination. In addition, the figure does not include any material combinations that are only represented to a limited extent in the literature, for example, FMLs with natural fibers or with metals such as magnesium. The relative distribution of the different material combinations in the remaining 1635 publications is presented in Fig. 2.2. The results show that more than half of all these publications deal with the combination of aluminum and GFRP. There is also great interest in FMLs containing CFRP. However, the combination of CFRP and aluminum, known as CARALL, shows challenges regarding the considerable difference in their electrochemical potential and resulting galvanic corrosion [110, 132]. The combination of CFRP with steel is less represented, although various potential applications have been found in the scientific literature.

One big difference between structural materials like GLARE, which is used for large-scale components, and local hybridization, which mainly focuses on areas responsible for transferring loads through bolts, is their MVF. The MVF is defined by [124]:

$$MVF = \frac{V_{\mathrm{m}}}{V_{\mathrm{lam}}} \quad \left(= \frac{\sum_{1}^{p} t_{\mathrm{m}}}{t_{\mathrm{lam}}}, \quad \text{for} \quad V_{\mathrm{m}}, V_{\mathrm{lam}} = \text{const.} \right) \tag{2.1}$$

[1] Search string: TITLE-ABS-KEY ("fib* metal laminate*")

Table 2.1 Coefficients of thermal expansion (longitudinal direction) of various materials used in FMLs for a temperature range of 0 °C to 100 °C [106].

Material	CTE in $1 \times 10^{-6}\,K^{-1}$
Aramid fiber	-2
Glass fiber	3.5 - 7.2
Carbon fiber	(-1.08) - (-0.455)
Epoxy resin	50-67
Aluminum	23.5
Steel	11.7
Titanium	8.6

with V_m and V_{lam} being the volume fractions of metal and the total laminate, p the number of metal plies, and t_m and t_{lam} the cross-sectional thicknesses of metal sheets and the total laminate. The simplified calculation with the thickness ratios only applies to areas with a constant volume fraction.

For local hybridization, the MVF is often gradually increased over a certain length to reduce stiffness discontinuities that might act as locations for crack initiation [89]. This leads to locally high MVF differences in this type of FML application. In function integration, the main objective is to utilize another advantage of the metal, like electrical conductivity, abrasion resistance, or energy dissipation, besides its load-bearing function. This is why MVFs are generally lower, e.g. 2% to 4% in [28]. For applications where the FML is being used as a structural material, the MVF very much depends on the density of the metal constituent. When metals like steel with a high density are used, the MVF again is rather small, e.g. 1% to 7% in [115]. When lightweight metals like aluminum are used, generally the metal sheet thicknesses and MVFs are higher, like in GLARE, e.g. 54% to 70% in [124], or in CARALL, e.g. 50% to 65% in [23].

2.2 Residual Stresses in Fiber Metal Laminates

Residual stresses exist within a material without any external forces being applied to it. They evolve whenever a part of the material is hindered in its free strain by internal or external boundaries [138]. Residual stresses may be purposely introduced into a material, e.g., to improve the damage tolerance of brittle materials like glass and ceramic or harden metal surfaces. On the other hand, mechanical stresses result from external forces and loads acting on a body. From a mechanical perspective, both types of stresses affect the material performance in the same way [46].

In most cases, residual stresses in composite materials are undesirable as they can decrease the load-bearing capacity [66], cause premature fatigue [5, 54], and result in geometric distortions [2]. In composite materials, residual stresses exist on different scales. On the micro-scale, residual stresses develop between the fibers and matrix (intralaminar), whereas, on the macro-scale, residual stresses evolve between different layers (interlaminar) of the composite due to their anisotropy [67, 138].

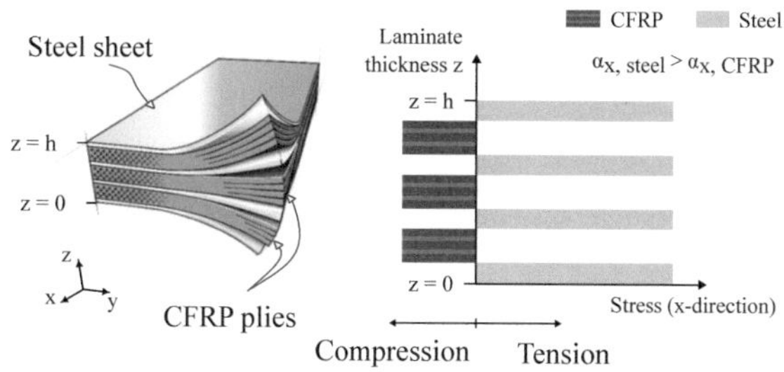

Fig. 2.3 A typical residual stress state in an FML with uni-directionally oriented fiber layers consists of tensile stresses in the metal sheets and compressive stresses in the fiber layers in fiber direction (adapted from [129]).

The main reasons for the evolution of residual stresses in composites on micro- and macro-scale are the incompatibility of the thermal expansion properties (Table 2.1) and the resin shrinkage during cure. Therefore, they are often referred to as curing stresses [86] or thermal (residual) stresses [59].

Although the residual stress state within a laminate may vary at different locations, the residual stresses σ_x over a laminate cross-section must be balanced. Therefore, for any cross-section perpendicular to the in-plane direction x, a force equilibrium exists [46]:

$$\int \sigma_x dz = 0 \tag{2.2}$$

with z defining the laminate thickness coordinate. Besides the force equilibrium, a balance of momentum exists in such an FML [46]:

$$\int \sigma_x z dz = 0 \tag{2.3}$$

This explains why laminates distort when an asymmetric residual stress state is introduced during processing. Forced asymmetry in the layer structure and consequently in the residual stress state concerning the laminate midplane can be used to determine the residual stresses generated during manufacturing [19, 136].

In FML, the interlaminar residual stresses are of primary importance. The main contribution to these stresses is the differences in the CTEs between fiber layers and metal sheets (Table 2.1), combined with high material stiffness and process temperatures during manufacturing [2, 12]. The contributions from the chemical shrinkage of the resin are small due to the high stiffness of the metals and fibers in the fiber direction.

Considering an FML with uni-directionally oriented fiber layers, the highest residual stresses generally developed in the fiber direction. The main reason is the lower CTEs of the fibers compared to the metals (Table 2.1), which leads to a restriction of the metal shrinkage during final cool-down from an elevated temperature level (typically 120 °C to 180 °C) in the manufacturing process of thermoset resins. At room temperature, the residual stress state consists of tensile stresses in the metal sheets and compressive stresses in the fiber layers in the fiber direction. Figure 2.3 illustrates such a stress state in an FML consisting of CFRP and steel.

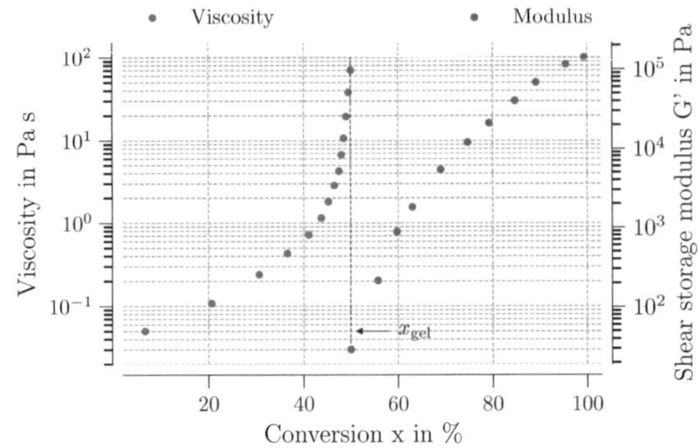

Fig. 2.4 After the conversion of the epoxy groups exceeds the gel point (x_{gel}), the mechanical behavior changes from viscous to visco-elastic (adapted from [122]).

All the influencing variables contributing to the residual stress state in any composite or FML can be categorized as intrinsic or extrinsic variables [4]. The intrinsic variables are driven by the material and layup, like material properties, ply stacking, and part geometry. Extrinsic parameters include every influence the part is subjected to during processing, like the local temperature and pressure over time, the tool material and surface condition, and the vacuum bagging [53, 119].

2.2.1 Mechanical Property Evolution of an Epoxy Resin During Cure

This work focuses on thermoset-based FMLs since most FRPs used in the aerospace industry are based on thermoset resins [52], and the most prominent FML application in the aerospace industry is GLARE consisting of an FRP with a thermoset matrix [5]. However, there is also research on thermoplastic-based FMLs [1,82,100] and residual stresses are also an issue here [81].

Understanding the role of the resin of thermoset-based FRPs is essential to comprehend the formation of residual stress in composites and FMLs. The thermosetting resin of an FRP polymerizes during processing so that a cross-linked polymer stabilized by covalent bonds is created from the number of monomers held together by Van der Waals forces. [122]

There are two significant points during polymerization where the mechanical polymer characteristics abruptly change. The first is the gel point (x_{gel}), defined by the amount of chemical conversion that correlates to the amount of monomers already reacted during polymerization. According to statistical considerations, the gel point is reached when a macro polymer extends to all system boundaries [32]. The resin abruptly changes from liquid to visco-elastic or rubbery at the gel point. Hence, the resin viscosity tends towards infinity at gelation, and a solid fraction with increasing elastic modulus becomes measurable, as illustrated in Fig. 2.4. With the beginning of the elastic behavior, stresses cannot be directly relaxed through viscous flow. The gel point is a constant material property not dependent on temper-

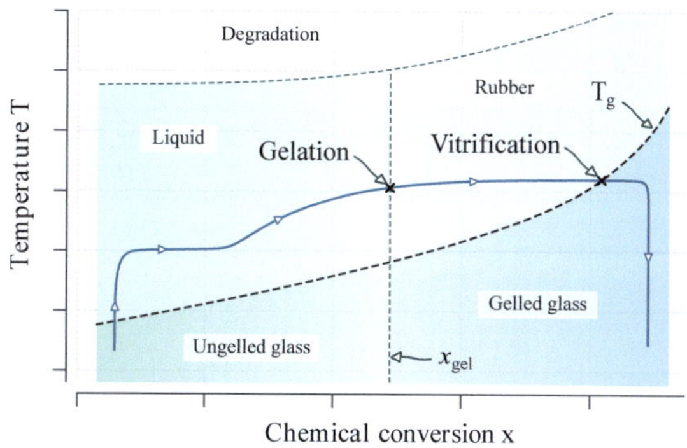

Fig. 2.5 Resin states illustrated in a conversion-temperature-transformation diagram for an epoxy resin (adapted from [122]) and the typical path followed during a two-hold curing process.

ature. It can be measured through thermo-mechanical tests like dynamic mechanical analysis (DMA). In contrast, detection in differential scanning calorimetry (DSC) is challenging since the cure rate does not change during gelation. [32, 38, 122]

The other significant point during cure is vitrification, which is reached when the instantaneous glass transition temperature of the resin surpasses the actual resin temperature (see Fig. 2.5). At vitrification, the resin changes its state from visco-elastic (rubber) to elastic (gelled glass) and possesses significant stiffness. Vitrification is reversible by increasing the temperature above the glass transition temperature. After vitrification, the curing rate is significantly reduced and becomes controlled by diffusion. Vitrification is detectable with DMA and DSC since the resin heat capacity significantly changes with vitrification. [38, 51, 130]

Figure 2.5 shows the typical path through the different stages a thermoset resin follows during manufacturing with a two-hold curing process. Initially, the material is stored below the initial glass transition temperature of the resin to prevent curing reactions before manufacturing. After thawing, the material is processed and subjected to heat in an oven or autoclave, leading to increased chemical conversion at elevated temperatures until gelation and vitrification occur. After cooling down from the manufacturing temperature, the material is in its glassy state (gelled glass).

Due to its viscous material behavior, the resin cannot transfer any loads in its liquid state. In this stage, the only stresses that might be introduced into a composite or FML result from interlocking between fibers or metal sheets and tools, which are independent of the resin.

From the gel point onwards, the resin starts to gain a measurable shear modulus (cf. Fig. 2.4) and can sustain or transfer certain amounts of loads. However, due to the visco-elasticity, the stresses can be partly relaxed. Again, stresses can also arise in this stage due to tool-fiber or tool-metal sheet interaction. After vitrification, the resin behaves purely elastic; the main stress contribution in this stage is thermal stresses due to differences in coefficients of expansion of the composite constituents. Any stresses introduced by the tool after vitrification are completely relaxed upon demolding, and therefore, there is no tool contribution to any residual stresses after this significant point during cure [51, 130].

Table 2.2: Different approaches for T_{sf} determination and consequent T_{sf} for the Hexcel 8552-AS4 prepreg material based on the technical data sheet [44] and values reported in [83].

$T_{sf} = T_{cure}$ [3]	$T_{sf} = T_g$ [12]	$T_{sf} = T_{g,\,onset} - 20\,°C$ [106]	$T_{sf} = \frac{2}{3} T_g$ [12]	
180	200 to 204	180 to 184*	133 to 136	°C

* $T_{g,\,onset}$ not available, therefore T_g is used instead ($T_{g,\,onset} < T_g$).

Consequently, to reduce residual stresses in a thermoset-based composite or FML at room temperature, the goal is to vitrify the resin as early as possible, at the lowest temperature possible, and at the maximum conversion possible. The reason is that stresses introduced in the viscous state are not stored in the material, and stresses introduced after vitrification are completely relaxed upon cool-down [3]. Consequently, for the investigation of the residual stress state in a composite, the most relevant phase during curing is the time between gelation and vitrification, excluding any contributions directly introduced into fiber or metal materials.

2.2.2 Stress-Free Temperature

The stress-free temperature of a laminate is important for the analytical calculation and the determination of the residual stress state. The stress-free temperature T_{sf} is a laminate-specific parameter and designates that temperature at which a laminate is considered to be in its stress-free state [19, 86]. This means that any laminate can be heated (or cooled) to a temperature until the residual stresses in each layer of the laminate, as shown in Fig. 2.3, become zero. The stress-free temperature is not only dependent on the resin state of cure but also sensitive to extrinsic variables influencing the residual stress state like cure cycle or part-tool interaction.

Determining T_{sf} is critical for accurate residual stress calculation. The literature [12, 106] proposes different theoretical approaches however, depending on the approach, the assumed stress-free temperature can deviate by as much as 71 K for a 180 °C curing epoxy resin, as shown in Table 2.2. Therefore, an experimental determination of the stress-free temperature is unavoidable for accurate residual stress predictions. The approximate methods in Table 2.2 assume a purely intrinsic determination of the stress-free temperature and, therefore, define it as a resin characteristic. Extrinsic influencing variables are neglected.

The T_{sf} can be experimentally determined using laminates with an asymmetric layup concerning the laminate midplane cured on a flat tool. The laminates start to curve upon demolding and provide the characteristic T_{sf} when reheated. This method was already used for both monolithic composites [19] and FMLs [95]. From these asymmetric specimens, also the curvature can be evaluated, and a corresponding T_{sf} can be calculated using bi-material beam theory [96, 118].

2.2.3 Analytic Description of Effective Ply Stresses

The interlaminar residual stresses and stresses due to external loads contribute to the single-ply stresses in an FML. The effective ply stress is a superposition of these two stress contributions [45]. With classical laminate theory (CLT), the effective ply stress $\boldsymbol{\sigma}_k$ in the single layers of an FML can be calculated [84, 106].

First, the plane stress stiffness matrix $\boldsymbol{Q}_k$ in material coordinates for each ply k needs to be calculated using the elastic moduli E_1 and E_2, the shear modulus G_{12} and the corresponding Poisson's ratios ν_{12} and ν_{21}

$$\boldsymbol{Q}_k = \begin{bmatrix} \frac{E_1}{1-\nu_{12}\nu_{21}} & \frac{\nu_{12}E_2}{1-\nu_{12}\nu_{21}} & 0 \\ \frac{\nu_{12}E_2}{1-\nu_{12}\nu_{21}} & \frac{E_2}{1-\nu_{12}\nu_{21}} & 0 \\ 0 & 0 & G_{12} \end{bmatrix} \tag{2.4}$$

where in the following, the material coordinate system is denoted with indices 1, 2, 3, and the laminate coordinate system with x, y, z.

With the transformation matrix $\boldsymbol{T}$

$$\boldsymbol{T} = \begin{bmatrix} c^2 & s^2 & 2sc \\ s^2 & c^2 & -2sc \\ -sc & sc & c^2 - s^2 \end{bmatrix} \tag{2.5}$$

where the terms s and c indicate $sin(\theta)$ and $cos(\theta)$, with θ being the ply angle in the laminate coordinates, the plane stress stiffness matrix can be transformed into the global coordinate system

$$\overline{\boldsymbol{Q}}_k = \boldsymbol{T}^{-1}\boldsymbol{Q}_k(\boldsymbol{T}^{-1})^T \tag{2.6}$$

and the $\boldsymbol{ABD}$ matrices of the laminate are calculated using the respective ply coordinates z

$$\boldsymbol{A} = \sum_{k=1}^{n} \overline{\boldsymbol{Q}}_k(z_k - z_{k-1}) \tag{2.7}$$

$$\boldsymbol{B} = \frac{1}{2}\sum_{k=1}^{n} \overline{\boldsymbol{Q}}_k(z_k^2 - z_{k-1}^2) \tag{2.8}$$

$$\boldsymbol{D} = \frac{1}{3}\sum_{k=1}^{n} \overline{\boldsymbol{Q}}_k(z_k^3 - z_{k-1}^3) \tag{2.9}$$

To incorporate the thermal expansion behavior for each ply k the corresponding coefficients of thermal expansion $\overline{\boldsymbol{\alpha}}_k$ for each ply k in the global coordinate system can be derived from the local coefficients of thermal expansion $\boldsymbol{\alpha}_k$

$$\overline{\boldsymbol{\alpha}}_k = \boldsymbol{T}^T\boldsymbol{\alpha}_k \tag{2.10}$$

The equivalent thermal forces $\boldsymbol{N}_t$ and moments $\boldsymbol{M}_t$ per unit length can subsequently be calculated

$$\boldsymbol{N}_t = \Delta T \sum_{k=1}^{n} \overline{\boldsymbol{Q}}_k \overline{\boldsymbol{\alpha}}_k (z_k - z_{k-1}) \tag{2.11}$$

$$\boldsymbol{M}_t = \frac{1}{2} \Delta T \sum_{k=1}^{n} \overline{\boldsymbol{Q}}_k \overline{\boldsymbol{\alpha}}_k (z_k^2 - z_{k-1}^2) \tag{2.12}$$

Consequently, the midplane strains $\boldsymbol{\varepsilon}^0$ and curvatures $\boldsymbol{\kappa}$ are derived

$$\begin{pmatrix} \boldsymbol{\varepsilon}^0 \\ \boldsymbol{\kappa} \end{pmatrix} = \begin{bmatrix} \boldsymbol{a} & \boldsymbol{b} \\ \boldsymbol{b}^T & \boldsymbol{d} \end{bmatrix} \left(\begin{pmatrix} \boldsymbol{N} \\ \boldsymbol{M} \end{pmatrix} + \begin{pmatrix} \boldsymbol{N}_t \\ \boldsymbol{M}_t \end{pmatrix} \right) \tag{2.13}$$

where $\boldsymbol{a}$, $\boldsymbol{b}$ and $\boldsymbol{d}$ indicate the inverse of the entire matrix consisting of the $\boldsymbol{ABD}$ submatrices.

The total strain in global coordinate system $\overline{\boldsymbol{\varepsilon}}_k$ for each ply is consequently defined by

$$\overline{\boldsymbol{\varepsilon}}_k = \boldsymbol{\varepsilon}^0 + z_k \boldsymbol{\kappa} \tag{2.14}$$

and transforming that total strain into the material coordinate system for each ply

$$\boldsymbol{\varepsilon}_k = (\boldsymbol{T}^T)^{-1} \overline{\boldsymbol{\varepsilon}}_k \tag{2.15}$$

the effective material strain $\boldsymbol{\varepsilon}_k^m$, consisting of mechanical and thermal portion, is calculated by subtracting the free thermal strain of the ply

$$\boldsymbol{\varepsilon}_k^m = \boldsymbol{\varepsilon}_k - \overline{\boldsymbol{\alpha}}_k \, \Delta T \tag{2.16}$$

and ultimately the effective ply stress $\boldsymbol{\sigma}_k^m$ in the material system is given by

$$\boldsymbol{\sigma}_k^m = \boldsymbol{Q}_k \, \boldsymbol{\varepsilon}_k^m \tag{2.17}$$

The temperature difference ΔT to account for the thermally induced stresses is based on the stress-free temperature T_{sf} and a reference temperature T_{r} at which the laminate stress level is of interest

$$\Delta T = T_{\mathrm{r}} - T_{\mathrm{sf}} \tag{2.18}$$

With the analytic solution above, the residual stresses in different types of FMLs are calculated as a function of the metal volume fraction and shown in Fig. 2.6. For all types of FMLs, the same stress-free temperature (180 °C) is assumed. The data shows that the tensile residual stresses in the metal sheets are highest when combining CFRP with steel in laminates with low MVFs. In reality, the residual stress in GLARE will be even lower since the GFRP material used in industrial GLARE is a 120 °C curing epoxy resin. Therefore, the relevant temperature difference in Eq. (2.18) is even smaller than assumed in this figure. The curves in Fig. 2.6 further show that the amount of residual stresses depends not only on the CTE difference between fiber

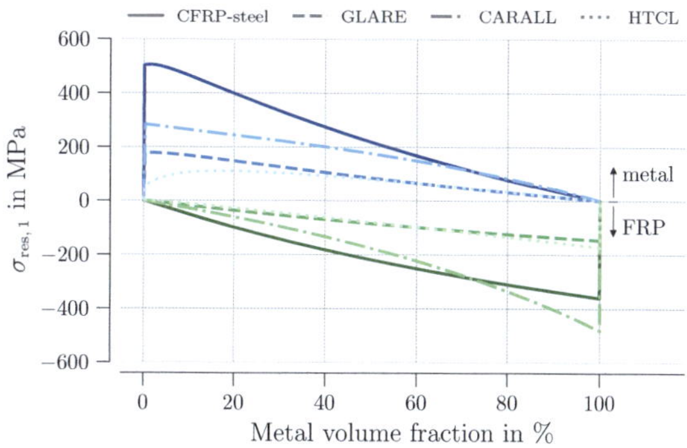

Fig. 2.6 Calculated residual stress state at room temperature as a function of the metal volume fraction in the primary laminate direction (fiber direction) for unidirectionally oriented FMLs of different types and an assumed stress-free temperature of 180 °C. The colors differentiate between the stresses in the metal sheets (blue) and FRP layers (green).

and metal material but also on the stiffness ratio of the two constituents, which is indicated by the non-linear characteristic of the curves.

2.2.4 Modification of Residual Stress States

Residual stresses are generally undesired because they reduce the effective load-carrying capacity. Therefore, different methods have been developed to manipulate the residual stress state in FMLs.

One way to reduce residual stresses is to pre-stretch the fiber layers to force them to shrink by the same amount as the metal during the final cool-down in the cure cycle. This can be achieved by clamping the edges of the fiber layers in the fiber direction with a clamping device with the same thermal expansion coefficient as the metal used in the FML [40, 133]. This ensures that they undergo the same pre-expansion as the metal during the heating steps in the cure cycle and consequently shrink by the same amount in the final step of the cure cycle [94, 116]. While this approach is effective, the applicability is restricted to limited part complexity and complicated for laminates with fibers oriented in several directions.

Another possibility to change the residual stress state after manufacturing is to post-stretch the laminates so that the metal sheets in the FML are plastically deformed. By doing so, the typical residual stress state (cf. Fig. 2.3) in an FML can be reversed with compressive stresses in the metal sheets and tensile stresses in the fiber sheets [5]. However, during post-stretching, the favorable stress state is only applied in one direction, which is inconvenient for FML components designed for more than uniaxial loading conditions. Post-stretching is mainly applicable for planar structures, and it is difficult to ensure a uniform stress distribution during post-stretching, leading to a significant variation in the residual stress state [5, 121]. For ARALL laminates, it was shown that post-stretching is even reducing the impact resistance since the fibers are in a pre-stretched state and cannot take as much load as in the unstretched state [103, 123].

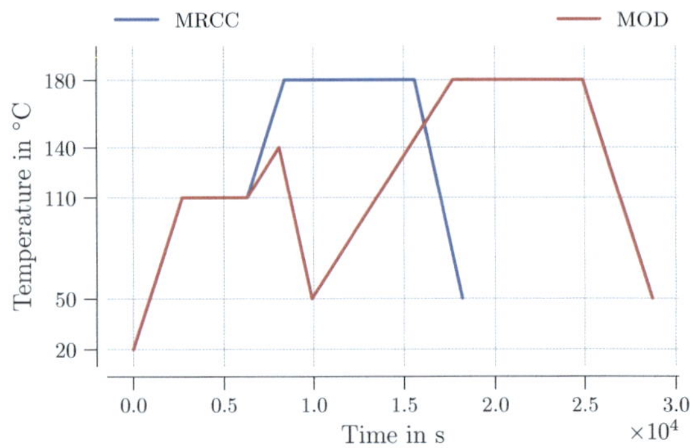

Fig. 2.7 Comparison of a modified cure cycle developed by Prussak et al. [96] for the Hexcel 8552-AS4 prepreg material with the corresponding manufacturer-recommended cure cycle .

A more promising approach to changing residual stresses is using so-called smart or modified cure cycles that have been investigated for composites [62, 128] and hybrid materials like FMLs [31,59,96]. In contrast to post-stretching, where the residual stresses are reversed, cure cycle modifications can only reduce residual stress inside a laminate. Thus, the leverage of this method is much smaller. However, the method is applicable for more complex layups, part geometries, and loading conditions and, therefore, a promising tool to increase the performance of FML structures. Modified cure cycles are designed such that the point at which the individual layers of the laminate start to transfer loads during manufacturing is shifted to the lowest temperature possible by adjusting the temperature cure profile [96]. This significantly reduces internal stresses in the laminate within the operating temperature range [24, 62, 137].

Figure 2.7 illustrates the adjustment in the modified (MOD) cure cycle and compares it to the manufacturer-recommended cure cycle (MRCC). The modified cure cycle for the Hexcel 8552-AS4 prepreg material, developed by Prussak et al. [96], incorporates an intermediate cooling step after a temperature of 140 °C is reached. This cooling step causes the resin to pass the gel point at a lower temperature during the subsequent cooling step and effectively reduces the residual stress state of the laminate.

The modification of cure cycles increases the quasi-static tensile strength of FRP [16, 57, 60] and FML [59, 94]. Besides the quasi-static strength, the fatigue strength of an FML can also be affected by the residual stress state [58], and cure cycle modifications can enhance the interfacial lap-shear strength of FML [56]. It was further shown that MOD cycles can improve the shear strength of aramid-fiber-reinforced bonded joints [18], and the strength of adhesive joints under cryogenic temperatures [70]. Furthermore, modifying cure cycles can help to reduce the residual stress inhomogeneity in thick composite laminates [61] and decrease the distortions in multilayer composites [112]. Moreover, adjusting the temperature profile during the cure of thin-walled monolithic composite profiles can influence the stability under axial compression [10, 21, 26].

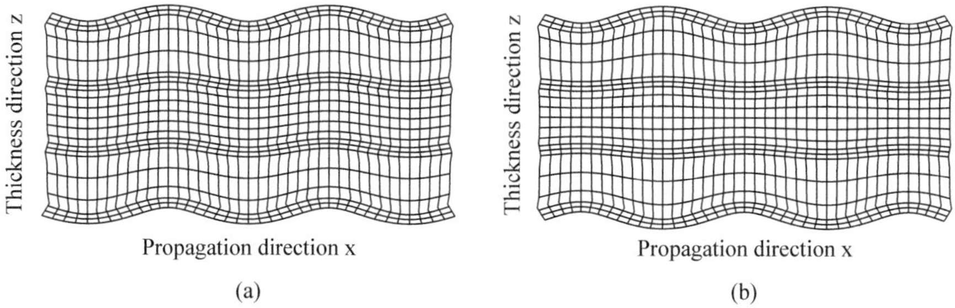

Fig. 2.8: Mode shapes of the two fundamental GUW wave modes with particle motions in x- and z-direction in the x-z-plane: **(a)** asymmetric A_0-mode and **(b)** symmetric S_0-mode.

2.3 Guided Ultrasonic Wave-Based Structural Health Monitoring of Fiber Metal Laminates

Following SAE International [104], SHM is one of the various methods employed for managing an aircraft's structural health. Other methods include crew observations, maintenance records, and fleet-wide data analysis. Although SHM focuses solely on the aircraft's structural components, health management is also used for engines and other aircraft systems. [33, 104]

Airliners, operators, and MRO (maintenance, repair, and overhaul) providers are motivated to use SHM systems to reduce maintenance costs and increase operational availability by reducing aircraft downtime. Therefore, SHM systems are designed to replace existing and time-consuming inspection procedures. The long-term goal is to integrate automated SHM systems into airborne structures, providing continuous information about structural health and allowing for condition-based maintenance (CBM) [33].

An SHM system generally can be used for several tasks, whereas the minimum requirement is to meet the first condition [101, 104]:

- Identify an event surpassing a certain threshold
- Determine the location of the event
- Assess the size or scale of the event
- Estimate the event's impact on the structural performance or residual life

To solve these tasks, many different sensor technologies have been developed for SHM applications [8, 113]. In these technologies, the use of guided ultrasonic waves (GUWs) has shown to be promising for aerospace structures made of composite materials [36, 37]. GUWs are very sensitive to damage inside a composite laminate. They can travel over large distances in thin structures with only small attenuation, making them ideal to cover large structures using little energy and sparse sensor networks [69].

GUWs emerge when the wave propagation medium is restricted and the different wave modes superpose. Some waves only propagate at the structure's surface (Rayleigh, Stonley, Scholte, and Love waves), which implies that only near-surface

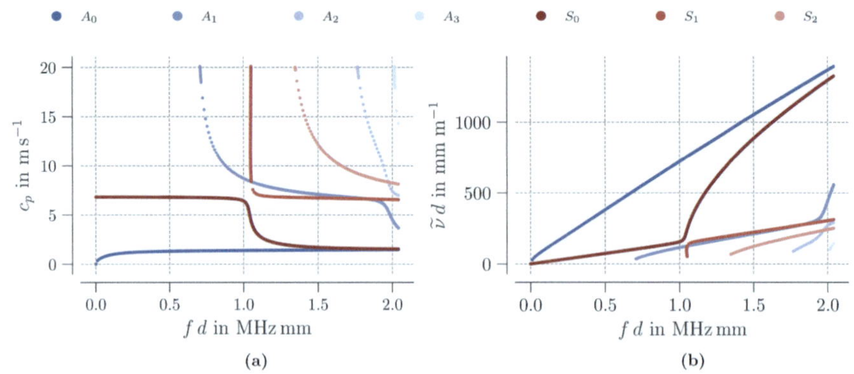

Fig. 2.9: Dispersion diagrams for a CFRP-steel laminate calculated with the open source software Dispersion Calculator [47] over a frequency-thickness range ($f\ d$) up to 2 MHz mm for the fundamental antisymmetric (A) and symmetric (S) wave modes: **(a)** in phase velocity (c_p) and **(b)** in wavenumber-thickness ($\tilde{\nu}\ d$) representation.

damage could be detected with these types of GUWs in an SHM system. The most common GUWs for SHM applications are Lamb waves. They result from the constructive superposition of longitudinal and flexural waves reflected and converted at the two parallel free surfaces in plate-like structures. Lamb waves are guided between the two surfaces and propagate within the plane of the plate. [105, 113]

At any given frequency, there are at least two modes that are either symmetrical or antisymmetrical concerning the particle motion relative to the center plane of the waveguide. At low frequencies, two fundamental wave modes exist: the antisymmetric A_0- and the symmetric S_0-mode. The particle motion of these mode shapes is illustrated in Fig. 2.8. At low frequencies, the A_0-mode can be approximated by a flexural wave, while the S_0-mode shows dominant longitudinal wave behavior. [22, 36]

Besides their multimodality, another characteristic of most GUWs is their dispersive behavior, which means that the wave modes' phase velocity depends on the frequency and the waveguide's thickness. This dispersive characteristic is illustrated in Fig. 2.9 in wavenumber thickness representation and regarding the phase velocity of the different Lamb wave modes. While the A_0-mode has a relatively constant phase-velocity over the frequency-thickness product, except for small frequency-thickness products, the S_0-mode shows a high dispersive characteristic at around 1 MHz mm for this material. The figure further shows that for increasing frequency-thickness products, the number of wave modes increases, whereas for small frequency-thickness products, only the two fundamental modes A_0 and S_0 exist. [36]

GUW-based SHM systems make use of the fact that the propagating waves interact with damage in a structure. The type of interaction depends on damage characteristics, such as the size of the damage concerning the wavelength, the position of the damage in the direction of the laminate thickness, or the type of damage (e.g., crack, delamination). The interaction mechanisms of GUWs with the damage range

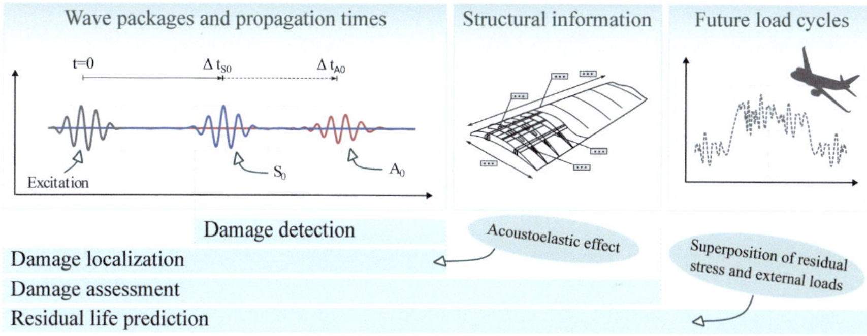

Fig. 2.10: Different types of information are needed to solve the four tasks of a GUW-based SHM system. The literature shows that stresses influence wave propagation velocity (acoustoelastic effect), and residual stresses superpose with external loads during fatigue. It is, therefore, assumed that residual stress states in FML are most likely to affect the two tasks of damage localization and residual life prediction.

from reflection, transmission, and absorption to velocity changes, scattering, and mode conversion. [105]

No matter which type of damage and wave propagation feature is used, GUW-based SHM is always an inverse problem. Therefore, a unique correlation between damage and wave propagation is necessary. It is crucial to know other parameters that influence the features used for damage detection, localization, and assessment to compensate for their influences [37, 111]. The literature shows that GUWs are sensitive to changing environmental and operational parameters like temperature [20, 25, 65], load or stress [13, 17, 79, 134]. The sensitivity of GUW towards stress seems especially relevant for SHM applications in FML, which contain an inherent residual stress state due to the manufacturing process.

2.3.1 Dependencies Between Residual Stress and Structural Health Monitoring

The motivation is to evaluate the relevance of the manufacturing-induced residual stress states for SHM with GUW. Therefore, the tasks of an SHM system most likely to be affected by the residual stress state in an FML need to be identified. To achieve this, the information required for each of the four SHM tasks is illustrated in Fig 2.10.

For damage detection, it is assumed that damage in a composite structure or an FML introduces an inhomogeneity considerably more significant than the interlaminar residual stress state itself. Therefore, the residual stress influence on any features of a wave packet used for damage detection is assumed to be of minor importance. Although Martinez et al. [76] indicate that the GUW response towards a fatigue crack is not only affected by the geometric crack dimension but also by the stress field around the cracked region. For impact damage, it was shown that residual stresses have a negligible influence on the damage characteristics [6, 80]. Equally, the results of [74, 77] indicate that damage detection in FML is independent of potential residual stress states in the laminate.

Damage localization, however, requires the propagation velocity of the GUW, especially for the most common methods using the time-of-flight (ToF) or delay-and-sum algorithms that rely on a known propagation speed of the evaluated wave packet [108, 109]. Hence, a small deviation in propagation speed due to a residual stress state will reduce the localization accuracy, especially for long propagation paths.

For damage assessment, the damage location and relevant wave packet features must be related to structural information. Stress is expected to influence the wave propagation velocity of GUWs mainly. It is unclear which other relevant GUW features might be affected by stress. Therefore, the focus will be on the propagation velocity for the time being since an influence from isotropic materials is known.

Meanwhile, for residual life predictions, the SHM system needs to extrapolate into the future to determine the expected structural loads. Since it is known that residual stresses superpose with external loads in FML [5], a significant influence of the residual stress state is expected. For GLARE, it is shown that changing the residual stress state by post-stretching directly influences the fatigue lifetime of a laminate.

Consequently, if a manufacturing-induced residual stress state in an FML influences an SHM system, the most considerable influence is expected to be in damage localization and residual life prediction. Therefore, this work focuses on the influence of residual stresses on these two tasks of an SHM system. The following two sections will discuss relevant literature sources showing the influence of residual stress on GUW propagation and fatigue life of FML.

2.3.2 Stress Influence on Guided Ultrasonic Wave Propagation

The velocity of ultrasonic waves shows a dependency on a stress state prevailing in the material. This dependence is utilized, for example, in determining residual stresses in metallic materials by measuring the ToF of an ultrasonic wave through the material [131]. The change in the speed of sound due to stress in an elastic body can be both positive and negative depending on the direction of propagation and polarization of the propagating wave [73]. Pohl et al. [90] show that the change in velocity highly depends on the material itself (structural steel S355 vs. stainless steel 1.4301) when considering the wave polarized and propagating perpendicular to the direction of stress.

The dependency between a wave's velocity and stress is generally known as the acoustoelastic effect. The classical theory of elasticity, which assumes infinitesimal deformations, a unique linear dependence between stress and strain, and the instantaneous adoption of the state of equilibrium in the event of a change in stress or strain, cannot describe the acoustoelastic effect [85]. Therefore, higher-order non-linear elastic theories have been proposed to describe the stress-strain dependence of non-linear elastic (hyperelastic) materials [48, 87]. While the developed theories consider general sound waves in elastic bodies, some researchers also set the focus on GUWs [49, 98] and residual stress [75, 107].

Based on the non-linear elastic theory, the acoustoelastic effect of GUW in layered media has several specific features. The change in wave velocity due to stress depends on the frequency. There are frequency ranges with a significant dependence between the velocity and stress, ranges with a minor dependence, and, for each mode, a particular frequency exists where the wave speed is independent of stress. For a constant frequency, the change in wave speed is linearly related to the applied stress. In general, the stress-dependency of the wave speed increases with higher modes. [7, 39]

Furthermore, the frequency dependence of the acoustoelastic effect can also be related to mass and displacement fractions involved in the longitudinal and transverse directions [99]. At low frequencies, the S_0-mode has a predominant proportion of oscillating mass in the longitudinal direction. It, therefore, behaves predominantly like a longitudinal wave, in which tensile stress reduces the wave speed in the direction of propagation [90]. Only with increasing frequency do the vibration components perpendicular to the direction of propagation play a significant role, and the dependence of the change in velocity on the tensile stress decreases. The opposite is true with the A_0-mode, approximated by a flexural wave at low frequencies. Due to the predominant out-of-plane component for displacement and the masses involved, the velocity at low frequencies increases with increasing tensile stress. With increasing frequency and increasing in-plane oscillation components, the change in velocity is reduced when tension is applied [99].

However, there are only a few sources in the experimental literature that support these theoretical considerations regarding the influence of stresses on GUWs, and they focus on isotropic materials [9, 34, 88, 97]. Therefore, it is unknown how a complex residual stress state in FMLs interacts with the different GUW modes and whether a corresponding change in wave velocity is significant for a structural health monitoring system.

2.3.3 Residual Stress Influence on FML Fatigue

Research on GLARE shows that fatigue crack growth can only be adequately explained by considering the interlaminar residual stress state [117]. To further quantify the influence in GLARE laminates, the previously mentioned method of post-stretching was used to reverse the unfavorable residual stress state [5]. Khan et al. [54] showed that the final compressive stresses in the metal layers lead to lower stress concentrations at a crack tip and, consequently, to extended fatigue life. Similar results were found for CARALL by Lin et al. [72], as well as for thermoplastic-based hybrid glass and carbon-fiber-reinforced aluminum laminates [82]. For CFRP-steel laminates, Both [12] uses theoretical considerations to show a potential load increase of 79% under cyclic loading. The influence of post-stretching on the fatigue crack and delamination growth characteristics is also numerically described by Yeh [135] for ARALL. For more complex GLARE specimens using stiffening doublers, Vlot et al. [125] revealed that post-stretching can lead to increased fatigue delamination behavior, thus reducing fatigue life.

The effect of the modified cycles and the corresponding reduced residual stress on the fatigue behavior has been little studied. Kim et al. [60] investigated the influence of residual stress on cross-ply 0°/90° FRP layups and showed that the fatigue life could be extended with modified cure cycles. An increase in fatigue life of up to 600% is stated for low cycle fatigue (LCF). Kim et al. [58] show that a modified cure cycle can increase the torsional fatigue life of a hybrid aluminum/composite tube for high cycle fatigue. In contrast to [60], however, the influence vanishes for load levels within the LCF regime. Khodja et al. [55] investigated the influence of cure cycle modifications on the fatigue performance of aluminum sheets with FRP repair patches. However, it remains unclear why increasing the curing time by 45 min at the recommended curing temperature of 120 °C shows the best fatigue results, although the highest component distortions indicate the highest residual stresses generated. Hausmann and Schmidt [42] applied an asymmetric cooling strategy in a hot press for asymmetric laminates (steel / CFRP) to reduce residual stress and showed that a 10% increase in cycles to failure in fatigue tests is achieved. Laliberté et al. [68] demonstrated that using modified cure cycles and consequently reduced residual stress levels in GLARE laminates can shift fatigue initiation (defined as a 2 mm crack) towards higher numbers of cycles. However, a reduced fatigue initiation life for low residual stress levels was observed for the highest load level tested. Ji et al. [50] reveal that a residual stress state even influences the fatigue life after impact in thermoplastic CFRP-titanium laminates. Stress concentrations at the crack tip after impact are superposed by the tensile residual stresses in the metal layers, which accelerates the propagation of fatigue cracks.

The literature review shows that there is substantial interest in reducing residual stresses in FMLs to increase the performance of FML structures. Modifying the temperature profile during curing seems the most promising approach for arbitrary laminates with fibers oriented in several directions for multidirectional load cases. However, the literature focusing on the fatigue of FMLs reveals inconsistencies in their experimental results, especially when comparing the results of [58] and [60] for the LCF regime and within the results of [68]. It is also unknown whether residual stresses in CFRP-steel laminates are relevant for residual life predictions using SHM systems.

2.4 Research Hypotheses and Thesis Outline

The lack of industrial SHM applications in aviation, despite its great potential, shows that the technology is not yet sufficiently mature and that further research is needed. Only a few niche applications in the aerospace industry are known. For example, the Norwegian military uses a piezo-electric sensor array attached to the aluminum gearbox of aged helicopters to detect fatigue cracks in their initiation phase [126]. Another example from Delta Airlines uses comparative vacuum monitoring sensors attached to the rear pressure bulkhead of a Boeing 737 to detect fatigue surface cracks around rivet holes. The application of the SHM technology prevents the galley from being removed for visual inspections, which means considerable time savings in maintenance [102].

In particular, the ability to sense damage inside a laminate with GUWs shows great potential to ensure the structural integrity of highly loaded FML components during operation due to the material's complex damage characteristics. However, the use of GUWs to fulfill the various tasks of an SHM system in FMLs has received little scientific attention. In particular, the characteristic of FMLs to form a complex interlaminar residual stress state during manufacturing must be addressed for such an SHM system. Therefore, this work aims to investigate the influence of different residual stress states on the tasks of an SHM system. Based on the current state of research, the most considerable influence of residual stresses is expected to occur regarding the wave propagation velocity and the fatigue behavior. Therefore, the two tasks of damage localization and residual life prediction of an SHM system were identified as particularly affected.

Four research hypotheses are formulated to advance the path for SHM systems in FMLs. The first two hypotheses result from the need to quantify residual stress states before their influence on other parameters can be evaluated. The third and fourth hypotheses are based on the state of research regarding SHM in FMLs, which raises several questions:

! Research hypotheses

1. *In contrast to FRPs, strain sensors attached to the metal layers of FMLs enable the recording of meaningful time-based strain data from the beginning of the manufacturing process.*

2. *Asymmetric specimens can determine the residual stress states in FMLs with any dimension and layer structure by deriving a parameter only sensitive to extrinsic manufacturing process variables.*

3. *Different cure cycle-induced residual stress states in an FML considerably influence the phase velocity of the fundamental GUW modes (A_0 and S_0) propagating in such laminates.*

4. *The reduction of cure cycle-induced residual stresses improves the fatigue life of CFRP-steel laminates under constant amplitude tension-tension fatigue loading.*

Some restrictions are made within the scope of the dissertation to investigate these hypotheses. Even though the motivation for SHM in FMLs applies in general,

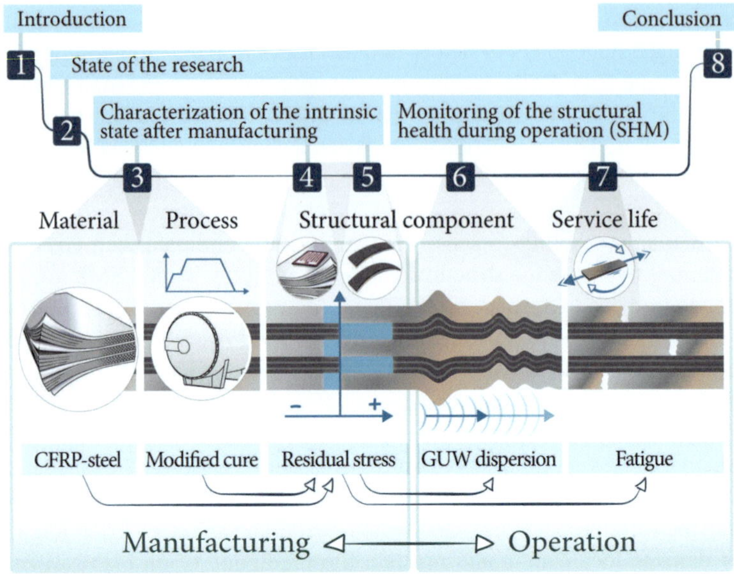

Fig. 2.11: This thesis is structured around key elements of the manufacturing and operation of an FML equipped with a GUW-based SHM system.

only FMLs consisting of CFRP and steel are considered in this work. The reason for selecting this material system is the wide range of applications on the one hand and the exceptionally high residual stresses on the other (cf. Fig. 2.6). In addition, modified curing cycles for this material system have already been presented in the literature (cf. Fig. 2.7), which can be referred to. It is also important to mention that while metals are part of the FMLs, this dissertation does not address the typical near-surface residual stresses associated with monolithic metal materials. These stresses exist in the metal sheets before the FML manufacturing process, but they remain unchanged throughout the FML manufacturing process. Furthermore, any influence of subsequent process steps in industrial manufacturing, like milling or drilling, and any stresses introduced by assembly processes are not addressed either. The focus is on the interlaminar residual stresses between the single layers of an FML on the macro-scale.

The formulated research hypotheses serve as the basis for the structure of this thesis, which is presented in Fig. 2.11. The motivation and the state of the research are introduced in Chaps. 1 and 2. The following chapters can be divided into the two main phases of a product lifecycle, manufacturing and operation. In the part addressing the manufacturing phase, Chap. 3 introduces all the relevant FML materials and manufacturing processes used during the investigations. In Chap. 4, a strain gauge method is presented to quantify the residual stress state in FMLs directly within the manufacturing process. Another method for residual stress quantification is discussed in Chap. 5 that uses asymmetric specimens and is better applicable to industrial manufacturing processes. With the knowledge of the residual stress states, the consequences of these states for SHM systems during the operation of FML components are investigated in Chaps. 6 and 7. While Chap. 6 discusses the influence of the residual stress state on the GUW propagation, Chap. 7 estimates the influence of the residual stresses on the laminate fatigue life. In Chap. 8, the hypotheses formulated in Chap. 2 are evaluated based on the results, and open questions are discussed.

References

1. Abdullah, M.R., Prawoto, Y., Cantwell, W.J.: Interfacial fracture of the fibre-metal laminates based on fibre reinforced thermoplastics. Materials & Design **66**, 446–452 (2015). doi:10.1016/j.matdes.2014.03.058
2. Abouhamzeh, M.: Distortions and residual stresses of GLARE induced by manufacturing. Ph.D. thesis, Delft University of Technology (2016). doi:10.4233/uuid:1f1b3e5c-72b8-440c-8d98-4b4c82814fb6
3. Adolf, D., Martin, J.E.: Calculation of Stresses in Crosslinking Polymers. Journal of Composite Materials **30**(1), 13–34 (1996). doi:10.1177/002199839603000102
4. Albert, C., Fernlund, G.: Spring-in and warpage of angled composite laminates. Composites Science and Technology **62**(14), 1895–1912 (2002). doi:10.1016/S0266-3538(02)00105-7
5. Alderliesten, R.: Fatigue and fracture of fibre metal laminates, *Solid mechanics and its applications*, vol. 236. Springer, Cham (2017)
6. Alderliesten, R.C.: Designing for damage tolerance in aerospace: A hybrid material technology. Materials & Design **66**, 421–428 (2015). doi:10.1016/j.matdes.2014.06.068
7. Babich, S.Y., Guz', A.N., Zhuk, A.P.: Elastic waves in bodies with initial stresses. Soviet Applied Mechanics **15**(4), 277–291 (1979). doi:10.1007/BF00884760
8. Balageas, D., Fritzen, C.P., Güemes, A.: Structural Health Monitoring. Wiley ISTE Ltd., London (2006)
9. Barth, T., Rauter, N., Lammering, R.: Experimental and numerical investigations on the acoustoelastic effect in hyperelastic waveguides. Structural Health Monitoring (2024). doi:10.1177/14759217241246506
10. Bienias, J., Gliszczynski, A., Jakubczak, P., Kubiak, T., Majerski, K.: Influence of autoclaving process parameters on the buckling and postbuckling behaviour of thin-walled channel section beams. Thin-Walled Structures **85**, 262–270 (2014). doi:10.1016/j.tws.2014.09.007
11. Boose, Y., Kappel, E., Stefaniak, D., Prussak, R., Pototzky, A., Weiß, L.: Phenomenological investigation on crash characteristics of thin layered CFRP-steel laminates. International Journal of Crashworthiness **182**(15), 1–10 (2020). doi:10.1080/13588265.2020.1787681
12. Both, J.C.: Tragfähigkeit von CFK-Metall-Laminaten unter mechanischer und thermischer Belastung: Zugl.: München, Techn. Univ., Diss., 2014. Ingenieurwissenschaften. Dr. Hut, München (2014). URL `http://mediatum.ub.tum.de/node?id=1173300`
13. Bulletti, A., Capineri, L.: Interdigital Piezopolymer Transducers for Time of Flight Measurements with Ultrasonic Lamb Waves on Carbon-Epoxy Composites under Pure Bending Stress. Journal of Sensors **2015**, 1–11 (2015). doi:10.1155/2015/259621
14. Burianek, D., Spearing, S.: Fatigue damage in titanium-graphite hybrid laminates. In: 39th AIAA/ASME/ASCE/AHS/ASC Structures, Structural Dynamics, and Materials Conference and Exhibit. American Institute of Aeronautics and Astronautics, Reston, Virigina (1998). doi:10.2514/6.1998-1959
15. Chai, G.B., Manikandan, P.: Low velocity impact response of fibre-metal laminates – A review. Composite Structures **107**, 363–381 (2014). doi:10.1016/j.compstruct.2013.08.003

16. Chava, S., Namilae, S., Al-Haik, M.: Residual stress reduction during composite manufacturing through cure modification: In situ analysis. Journal of Composite Materials p. 002199832110665 (2022). doi:10.1177/00219983211066545
17. Chen, F., Wilcox, P.D.: The effect of load on guided wave propagation. Ultrasonics **47**(1-4), 111–122 (2007). doi:10.1016/j.ultras.2007.08.003
18. Choi, I., Lee, D.G.: Surface modification of carbon fiber/epoxy composites with randomly oriented aramid fiber felt for adhesion strength enhancement. Composites Part A: Applied Science and Manufacturing **48**, 1–8 (2013). doi:10.1016/j.compositesa.2013.01.003
19. Crasto, A.S., Kim, R.Y.: On the Determination of Residual Stresses in Fiber-Reinforced Thermoset Composites. Journal of Reinforced Plastics and Composites **12**(5), 545–558 (1993). doi:10.1177/073168449301200505
20. Croxford, A.J., Moll, J., Wilcox, P.D., Michaels, J.E.: Efficient temperature compensation strategies for guided wave structural health monitoring. Ultrasonics **50**(4-5), 517–528 (2010). doi:10.1016/j.ultras.2009.11.002
21. Czapski, P., Jakubczak, P., Bieniaś, J., Urbaniak, M., Kubiak, T.: Influence of autoclaving process on the stability of thin-walled, composite columns with a square cross-section – Experimental and numerical studies. Composite Structures **250**, 112594 (2020). doi:10.1016/j.compstruct.2020.112594
22. Deutsch, V., Platte, M., Vogt, M.: Ultraschallprüfung. Springer Berlin Heidelberg, Berlin, Heidelberg (1997). doi:10.1007/978-3-642-59138-9
23. Dhaliwal, G.S., Newaz, G.M.: Effect of Layer Structure on Dynamic Response and Failure Characteristics of Carbon Fiber Reinforced Aluminum Laminates (CARALL). Journal of Dynamic Behavior of Materials **2**(3), 399–409 (2016). doi:10.1007/s40870-016-0075-1
24. Djokic, D., Johnston, A., Rogers, A., Lee-Sullivan, P., Mrad, N.: Residual stress development during the composite patch bonding process: measurement and modeling. Composites Part A: Applied Science and Manufacturing **33**(2), 277–288 (2002). doi:10.1016/S1359-835X(01)00083-5
25. Dodson, J.C., Inman, D.J.: Thermal sensitivity of Lamb waves for structural health monitoring applications. Ultrasonics **53**(3), 677–685 (2013). doi:10.1016/j.ultras.2012.10.007
26. Droździel, M., Podolak, P., Czapski, P., Zgórniak, P., Jakubczak, P.: Failure analysis of GFRP columns subjected to axial compression manufactured under various curing-process conditions. Composite Structures **262**, 113342 (2021). doi:10.1016/j.compstruct.2020.113342
27. Düring, D., Petersen, E., Stefaniak, D., Hühne, C.: Damage resistance and low-velocity impact behaviour of hybrid composite laminates with multiple thin steel and elastomer layers. Composite Structures **238**, 111851 (2020). doi:10.1016/j.compstruct.2019.111851
28. Düring, D., Weiß, L., Stefaniak, D., Jordan, N., Hühne, C.: Low-velocity impact response of composite laminates with steel and elastomer protective layer. Composite Structures **134**, 18–26 (2015). doi:10.1016/j.compstruct.2015.08.001
29. Fan, J., Cantwell, W.J., Guan, Z.W.: The low-velocity impact response of fiber-metal laminates. Journal of Reinforced Plastics and Composites **30**(1), 26–35 (2011). doi:10.1177/0731684410386133
30. Fink, A.: Lokale Metallhybridisierung zur Effizienzsteigerung von Hochlastfügestellen in Faserverbundwerkstoffen. Dissertation, Technische Universität Carolo-Wilhelmina zu Braunschweig, Braunschweig (2010)
31. Fleischer, J. (ed.): Intrinsische Hybridverbunde für Leichtbautragstrukturen. Springer Berlin Heidelberg, Berlin, Heidelberg (2021). doi:10.1007/978-3-662-62833-1
32. Flory, P.J.: Molecular Size Distribution in Three Dimensional Polymers. I. Gelation 1. Journal of the American Chemical Society **63**(11), 3083–3090 (1941). doi:10.1021/ja01856a061
33. Foote, P.: New Guidelines for Implementation of Structural Health Monitoring in Aerospace Applications. SAE International Journal of Aerospace **6**(2), 525–533 (2013). doi:10.4271/2013-01-2219
34. Gandhi, N., Michaels, J.E., Lee, S.J.: Acoustoelastic Lamb wave propagation in biaxially stressed plates. The Journal of the Acoustical Society of America **132**(3), 1284 (2012). doi:10.1121/1.4740491

35. Gerendt, C., Dean, A., Mahrholz, T., Rolfes, R.: On the progressive failure simulation and experimental validation of fiber metal laminate bolted joints. Composite Structures **229**, 111368 (2019). doi:10.1016/j.compstruct.2019.111368
36. Giurgiutiu, V.: Structural health monitoring with piezoelectric wafer active sensors. Academic Press/Elsevier, Amsterdam (2008). URL `http://site.ebrary.com/lib/alltitles/docDetail.action?docID=10203616`
37. Giurgiutiu, V.: Structural health monitoring of aerospace composites. Academic Press is an imprint of Elsevier, London, UK (2015). URL `http://www.sciencedirect.com/science/book/9780124096059`
38. Gotro, J., Prime, R.B.: Thermosets. In: Encyclopedia of Polymer Science and Technology, pp. 1–75. John Wiley & Sons, Inc, Hoboken, NJ, USA (2017). doi:10.1002/0471440264.pst519.pub2
39. Guz, A.N.: Elastic Waves in Bodies with Initial (Residual) Stresses. International Applied Mechanics **38**(1), 23–59 (2002). doi:10.1023/A:1015379824503
40. Haga, O., Koyama, H., Kawada, K., Ogawa, K.: Residual Stress Control of Al/CFRP Hybrid Composite Material. Journal of the Society of Materials Science, Japan **41**(464), 728–733 (1992). doi:10.2472/jsms.41.728
41. Harhash, M., Fischer, T., Grubenmann, M., Hua, W., Heingärtner, J., Kuhtz, M., Gude, M., Hora, P., Ziegmann, G., Palkowski, H.: Top-hat crashboxes of thermoplastic fibre-metal-laminates processed in one-step thermoforming: Experimental and numerical study. Composites Part B: Engineering **226**, 109367 (2021). doi:10.1016/j.compositesb.2021.109367
42. Hausmann, J., Schmidt, S.: Thermal residual stresses in thermoplastic CFRP-steel laminates: Modification and influence on fatigue life. Journal of Micromechanics and Molecular Physics **06**(03), 49–55 (2021). doi:10.1142/S2424913021430050
43. Hengel, C., Heuts, C., Verpoorte, J.: Integrating Electromagnetic Functions into Fiber Metal Laminates – the Structural Challenge. In: I European Conference On Multifunctional Structures. CIMNE (2019). doi:10.23967/emus.2019.013
44. Hexcel Corporation: Product data sheet HexPly 8552 (2020). URL `https://www.hexcel.com/user_area/content_media/raw/HexPly_8552_eu_DataSheet.pdf`
45. Homan, J.J.: Fatigue initiation in fibre metal laminates. International Journal of Fatigue **28**(4), 366–374 (2006). doi:10.1016/j.ijfatigue.2005.07.030
46. Hosford, W.F.: Mechanical behavior of materials, second edition edn. Cambridge Univ. Press, Cambridge (2010)
47. Huber, A.: Numerical Modeling of Guided Waves in Anisotropic Composites with Application to Air-coupled Ultrasonic Inspection. Dissertation, Universität Augsburg (2021)
48. Hughes, D.S., Kelly, J.L.: Second-Order Elastic Deformation of Solids. Physical Review **92**(5), 1145–1149 (1953). doi:10.1103/PhysRev.92.1145
49. Husson, D.: A perturbation theory for the acoustoelastic effect of surface waves. Journal of Applied Physics **57**(5), 1562–1568 (1985). doi:10.1063/1.334471
50. Ji, C., Hu, J., Alderliesten, R., Yang, J., Zhou, Z., Sun, Y., Wang, B.: On the post-impact fatigue behavior and theoretical life prediction of CF/PEEK-titanium hybrid laminates using an energy dissipation approach. Composites Science and Technology **245**, 110354 (2024). doi:10.1016/j.compscitech.2023.110354
51. Johnston, A.A.: An integrated model of the development of process-induced deformation in autoclave processing of composite structures. Dissertation, University of British Columbia, Vancouver, Kanada (1997). doi:10.14288/1.0088805
52. Kappel, E.: Process Distortions in Composite Manufacturing - From an experimental characterization to a prediction approach for the global scale. Dissertation, Otto-von-Guericke University, Magdeburg (2013)
53. Kappel, E., Stefaniak, D., Fernlund, G.: Predicting process-induced distortions in composite manufacturing – A pheno-numerical simulation strategy. Composite Structures **120**, 98–106 (2015). doi:10.1016/j.compstruct.2014.09.069
54. Khan, S.U., Alderliesten, R.C., Benedictus, R.: Post-stretching induced stress redistribution in Fibre Metal Laminates for increased fatigue crack growth resistance. Composites Science and Technology **69**(3-4), 396–405 (2009). doi:10.1016/j.compscitech.2008.11.006

55. Khodja, M., Fekirini, H., Govender, G., Bachir Bouiadjra, B.: Effect of curing cycle on fatigue life of cracked AA7075-T6 aircraft sheet repaired with a boron/epoxy composite patch. Iranian Journal of Science and Technology, Transactions of Mechanical Engineering **46**(1), 85–97 (2022). doi:10.1007/s40997-020-00400-6
56. Kim, H.S., Lee, D.G.: Avoidance of fabricational thermal residual stresses in co-cure bonded metal-composite hybrid structures. Journal of Adhesion Science and Technology **20**(9), 959–979 (2006). doi:10.1163/156856106777657805
57. Kim, H.S., Lee, D.G.: Reduction of fabricational thermal residual stress of the hybrid co-cured structure using a dielectrometry. Composites Science and Technology **67**(1), 29–44 (2007). doi:10.1016/j.compscitech.2006.05.002
58. Kim, H.S., Park, S.W., Hwang, H.Y., Lee, D.G.: Effect of the smart cure cycle on the performance of the co-cured aluminum/composite hybrid shaft. Composite Structures **75**(1-4), 276–288 (2006). doi:10.1016/j.compstruct.2006.04.030
59. Kim, H.S., Park, S.W., Lee, D.G.: Smart cure cycle with cooling and reheating for co-cure bonded steel/carbon epoxy composite hybrid structures for reducing thermal residual stress. Composites Part A: Applied Science and Manufacturing **37**(10), 1708–1721 (2006). doi:10.1016/j.compositesa.2005.09.015
60. Kim, H.S., Yoo, S.H., Chang, S.H.: In situ monitoring of the strain evolution and curing reaction of composite laminates to reduce the thermal residual stress using FBG sensor and dielectrometry. Composites Part B: Engineering **44**(1), 446–452 (2013). doi:10.1016/j.compositesb.2012.04.021
61. Kim, J.W., Lee, J.H., Kim, H.G., Kim, H.S., Lee, D.G.: Reduction of residual stresses in thick-walled composite cylinders by smart cure cycle with cooling and reheating. Composite Structures **75**(1-4), 261–266 (2006). doi:10.1016/j.compstruct.2006.04.029
62. Kim, S.S., Murayama, H., Kageyama, K., Uzawa, K., Kanai, M.: Study on the curing process for carbon/epoxy composites to reduce thermal residual stress. Composites Part A: Applied Science and Manufacturing **43**(8), 1197–1202 (2012). doi:10.1016/j.compositesa.2012.02.023
63. Kolesnikov, B., Fink, A., Hühne, C., Stefaniak, D., Borgwardt, H.: Strukturelement aus einem Hybridlaminat (2010). URL `https://patents.google.com/patent/DE102010035324A1/de`
64. Kolesnikov, B., Wilmes, H., Herrmann, A., Pabsch, A.: Composite material with a reinforced connecting area (20.03.1999). URL `https://patents.google.com/patent/WO2000056541A1/en`
65. Konstantinidis, G., Drinkwater, B.W., Wilcox, P.D.: The temperature stability of guided wave structural health monitoring systems. Smart Materials and Structures **15**(4), 967–976 (2006). doi:10.1088/0964-1726/15/4/010
66. Koord, J.: Design methodology for highly loaded composite bolted joints with local metal hybridization at low temperature: DLR-Forschungsbericht - DLR-FB-2024-11. Dissertation, German Aerospace Center, Braunschweig (2024). doi:10.57676/mzxc-3158
67. Kropka, M.: Auswirkungen von interlaminaren Eigenspannungen auf die mechanischen Eigenschaften von thermoplastischen Hybridverbundwerkstoffen. Ph.D. thesis, University of Bayreuth (2020). doi:10.15495/EPub_UBT_00004636
68. Laliberté, J., Mahendran, M., Djokic, D., Li, C., Kratz, J.: Effect of process-induced residual stresses on mechanical properties and fatigue crack initiation in fibre metal laminates. In: L. Lazzeri (ed.) ICAF 2007 : durability and damage tolerance of aircraft structures: metals vs composites; proceedings of the 24th symposium of the International Committee on Aeronautical Fatigue, pp. 576–592. Pacini Ed., Pisa (2007)
69. Lammering, R., Gabbert, U., Sinapius, M., Schuster, T., Wierach, P. (eds.): Lamb-Wave Based Structural Health Monitoring in Polymer Composites. Research Topics in Aerospace. Springer International Publishing, Cham (2017). doi:10.1007/978-3-319-49715-0
70. Lee, K.H., Lee, D.G.: Smart cure cycles for the adhesive joint of composite structures at cryogenic temperatures. Composite Structures **86**(1-3), 37–44 (2008). doi:10.1016/j.compstruct.2008.03.018

71. Li, E.: Residual tension strength study of fatigued open-hole titanium-graphite hybrid composite laminates. In: 39th AIAA/ASME/ASCE/AHS/ASC Structures, Structural Dynamics, and Materials Conference and Exhibit. American Institute of Aeronautics and Astronautics, Reston, Virigina (1998). doi:10.2514/6.1998-1960
72. Lin, C.T., Kao, P.W., Yang, F.S.: Fatigue behaviour of carbon fibre-reinforced aluminium laminates. Composites **22**(2), 135–141 (1991). doi:10.1016/0010-4361(91)90672-4
73. Lüthi, T.: Beitrag zur Ermittlung von Eigenspannungen mit Ultraschall. Anwendung auf Schweissnähte
74. Maghsoodi, A., Ohadi, A., Sadighi, M., Amindavar, H.: Damage detection in multilayered fiber–metal laminates using guided-wave phased array. Journal of Mechanical Science and Technology **30**(5), 2113–2120 (2016). doi:10.1007/s12206-016-0418-9
75. Man, C.S., Lu, W.Y.: Towards an acoustoelastic theory for measurement of residual stress. Journal of Elasticity **17**(2), 159–182 (1987). doi:10.1007/BF00043022
76. Martinez, M., Pant, S., Yanishevsky, M., Backman, D.: Residual stress effects of a fatigue crack on guided lamb waves. Smart Materials and Structures **26**(11), 115004 (2017). doi:10.1088/1361-665X/aa8a0c
77. Masurkar, F.: Damage detection in hybrid metal-composite plates using ultrasonic guided waves based on outliers estimate. In: Structural Health Monitoring, Materials Research Proceedings, pp. 158–165. Materials Research Forum LLC (2023). doi:10.21741/9781644902455-20
78. Miller, J.L., Progar, D.J., Johnson, W.S., St.Clair, T.L.: Preliminary evaluation of hybrid titanium composite laminates: NASA Technical Memorandum 109095
79. Mohabuth, M., Kotousov, A., Ng, C.T.: Effect of uniaxial stress on the propagation of higher-order Lamb wave modes. International Journal of Non-Linear Mechanics **86**(9), 104–111 (2016). doi:10.1016/j.ijnonlinmec.2016.08.006
80. Moriniere, F.D.: Low-velocity impact on fibre-metal laminates. Dissertation, Delft University of Technology (2014). doi:10.4233/uuid:022e1a88-f48e-493e-805e-014f396a05dc
81. Mrzljak, S., Schmidt, S., Kohl, A., Hülsbusch, D., Hausmann, J., Walther, F.: Testing procedure for fatigue characterization of steel-CFRP hybrid laminate considering material dependent self-heating. Materials (Basel, Switzerland) **14**(12) (2021). doi:10.3390/ma14123394
82. Mrzljak, S., Trautmann, M., Wagner, G., Walther, F.: Very high cycle fatigue assessment of thermoplastic-based hybrid fiber metal laminate by using a high-frequency resonant testing system. International Journal of Fatigue p. 108361 (2024). doi:10.1016/j.ijfatigue.2024.108361
83. National Institute for Aviation Research - Wichita State University: Hexcel 8552 AS4 unidirectional prepreg at 190 gsm & 35% RC: Qualification material property data report (2011). URL `https://www.wichita.edu/research/NIAR/Research/hexcel-8552/AS4-Unitape-2.pdf`
84. Nettles, A.T.: Basic mechanics of laminated composite plates: Technical report: NASA-RP-1351 (1994). URL `https://ntrs.nasa.gov/citations/19950009349`
85. Ogden, R.W., Ogden, R.W.: Non-linear elastic deformations, 1. publ., unabridged and corr. republ edn. Dover books on physics. Dover Publ, Mineola, NY (1997). URL `http://www.loc.gov/catdir/description/dover031/97016162.html`
86. Pagano, N.J., Hahn, H.T.: Evaluation of Composite Curing Stresses. In: G.M.L. Gladwell, J.N. Reddy (eds.) Mechanics of Composite Materials, *Solid mechanics and its applications*, vol. 34, pp. 57–69. Springer Netherlands, Dordrecht (1994). URL `https://doi.org/10.1007/978-94-017-2233-9_6`
87. Pao, Y.H., Gamer, U.: Acoustoelastic waves in orthotropic media. The Journal of the Acoustical Society of America **77**(3), 806–812 (1985). doi:10.1121/1.392384
88. Pei, N., Bond, L.J.: Experimental study of the acoustoelastic Lamb wave in thin plates. AIP Conference Proceedings **1706**(1), 070009 (2016). doi:10.1063/1.4940527
89. Petersen, E., Stefaniak, D., Hühne, C.: Experimental investigation of load carrying mechanisms and failure phenomena in the transition zone of locally metal reinforced joining areas. Composite Structures **182**(6), 79–90 (2017). doi:10.1016/j.compstruct.2017.09.002

90. Pohl, J., Prautzsch, V., Holstein, P., Bodi, A., Pick, C.: Einflussgrößen bei der Bestimmung von Eigenspannungen mit Ultraschall. In: DGZfP Jahrestagung. NDT.net (2019). URL https://www.ndt.net/?id=24629
91. Pototzky, A.: Graphenunterstützte Konstruktionsmethode zur Funktionsintegration in Leichtbaustrukturen. Dissertation, Technische Universität Braunschweig (2020)
92. Pototzky, A., Stefaniak, D., Hühne, C.: Potentials of load carrying conductor tracks in new vehicle structures. In: K. Dröder, T. Vietor (eds.) Technologies for economical and functional lightweight design, Zukunftstechnologien für den multifunktionalen Leichtbau, pp. 79–90. Springer Berlin Heidelberg, Berlin, Heidelberg (2019). doi:10.1007/978-3-662-58206-0_8
93. Prasad, N.E., Wanhill, R.J.H.: Aerospace Materials and Material Technologies. Springer Singapore, Singapore (2017). doi:10.1007/978-981-10-2134-3
94. Prussak, R., Stefaniak, D., Hühne, C., Sinapius, M.: Residual stresses in intrinsic UD-CFRP-steel-laminates - Experimental determination, identification of sources, effects and modification approaches. In: 20th Symposium on Composites, Materials Science Forum, pp. 369–376. Trans Tech Pubn (2015). doi:10.4028/www.scientific.net/MSF.825-826.369. URL https://www.scientific.net/MSF.825-826.369.pdf
95. Prussak, R., Stefaniak, D., Hühne, C., Sinapius, M.: Evaluation of residual stress development in FRP-metal hybrids using fiber Bragg grating sensors. Production Engineering **12**(2), 259–267 (2018). doi:10.1007/s11740-018-0793-4
96. Prussak, R., Stefaniak, D., Kappel, E., Hühne, C., Sinapius, M.: Smart cure cycles for fiber metal laminates using embedded fiber Bragg grating sensors. Composite Structures **213**, 252–260 (2019). doi:10.1016/j.compstruct.2019.01.079
97. Qiu, L., Yan, X., Lin, X., Yuan, S.: Multiphysics simulation method of lamb wave propagation with piezoelectric transducers under load condition. Chinese Journal of Aeronautics **32**(5), 1071–1086 (2019). doi:10.1016/j.cja.2019.02.007
98. Qu, J., Liu, G.: Effects of Residual Stress on Guided Waves in Layered Media. In: D.O. Thompson, D.E. Chimenti (eds.) Review of Progress in Quantitative Nondestructive Evaluation, vol. 45, pp. 1635–1642. Springer US, Boston, MA (1998). doi:10.1007/978-1-4615-5339-7_212
99. Rauter, N., Barth, T., Lammering, R.: Numerical investigation on the relationship between the acoustoelastic effect and the modal properties of Lamb waves in aluminum. In: D. Saravanos, A. Benjeddou (eds.) 10th ECCOMAS Thematic Conference on Smart Structures and Materials, pp. 83–94. Dept. of Mechanical Engineering & Aeronautics University of Patras, Patras (2023). doi:10.7712/150123.9767.444077
100. Reyes, G., Kang, H.: Mechanical behavior of lightweight thermoplastic fiber–metal laminates. Journal of Materials Processing Technology **186**(1-3), 284–290 (2007). doi:10.1016/j.jmatprotec.2006.12.050
101. Ricci, F., Monaco, E., Boffa, N.D., Maio, L., Memmolo, V.: Guided waves for structural health monitoring in composites: A review and implementation strategies. Progress in Aerospace Sciences **129**, 100790 (2022). doi:10.1016/j.paerosci.2021.100790
102. Roach, D., Lynch-Staunton, T., Shaigec, B., Formosa, D.: Assessing the Performance CVM Sensors for Monitoring the 737 Aft Pressure Bulkhead. In: S. Farhangdoust, A. Guemes, F.K. Chang (eds.) Structural Health Monitoring 2023. Destech Publications, Stanford, CA (2023)
103. Sadighi, M., Alderliesten, R.C., Benedictus, R.: Impact resistance of fiber-metal laminates: A review. International Journal of Impact Engineering **49**, 77–90 (2012). doi:10.1016/j.ijimpeng.2012.05.006
104. SAE International: Guidelines for Implementation of Structural Health Monitoring on Fixed Wing Aircraft. doi:10.4271/ARP6461A
105. Schmidt, D.: Modenselektive Übertragung von Lambwellen in Faserverbundstrukturen. Dissertation, Technische Universität Braunschweig (2014). URL https://elib.dlr.de/90214/
106. Schürmann, H.: Konstruieren mit Faser-Kunststoff-Verbunden. Springer Berlin Heidelberg, Berlin, Heidelberg (2007). doi:10.1007/978-3-540-72190-1
107. Shams, M., Destrade, M., Ogden, R.W.: Initial stresses in elastic solids: Constitutive laws and acoustoelasticity. Wave Motion **48**(7), 552–567 (2011). doi:10.1016/j.wavemoti.2011.04.004

108. Shan, S., Qiu, J., Zhang, C., Ji, H., Cheng, L.: Multi-damage localization on large complex structures through an extended delay-and-sum based method. Structural Health Monitoring **15**(1), 50–64 (2016). doi:10.1177/1475921715623358
109. Sharif-Khodaei, Z., Aliabadi, M.H.: Assessment of delay-and-sum algorithms for damage detection in aluminium and composite plates. Smart Materials and Structures **23**(7), 075007 (2014). doi:10.1088/0964-1726/23/7/075007
110. Sinmazçelik, T., Avcu, E., Bora, M.Ö., Çoban, O.: A review: Fibre metal laminates, background, bonding types and applied test methods. Materials & Design **32**(7), 3671–3685 (2011). doi:10.1016/j.matdes.2011.03.011
111. Sohn, H.: Effects of environmental and operational variability on structural health monitoring. Philosophical transactions. Series A, Mathematical, physical, and engineering sciences **365**(1851), 539–560 (2007). doi:10.1098/rsta.2006.1935
112. Soohyun, N., Dongyoung, L., Ilbeom, C., Dai Gil, L.: Smart cure cycle for reducing the thermal residual stress of a co-cured E-glass/carbon/epoxy composite structure for a vanadium redox flow battery. Composite Structures **120**, 107–116 (2015). doi:10.1016/j.compstruct.2014.09.037
113. Staszewski, W.J. (ed.): Health monitoring of aerospace structures: Smart sensor technologies and signal processing. John Wiley, Chichester (2004). doi:57607
114. Stefaniak, D.: Improving residual strength of unidirectionally reinforced plastic laminates by metal layering. Dissertation, Technische Universität Carola-Wilhelmina zu Braunschweig (2017)
115. Stefaniak, D., Kappel, E., Kolesnikov, B., Hühne, C.: Improving the mechanical performance of unidirectional CFRP by metal-hybridization. In: ECCM15 - 15th European Conference on Composite Materials (2012)
116. Stefaniak, D., Kappel, E., Kolotylo, M., Hühne, C.: Experimental identification of sources and mechanisms inducing residual stresses in multi-layered Fiber-Metal- Laminates. In: Proceedings of Euro Hybrid Materials and Structures 2014 (2014)
117. Takamatsu, T., Matsumura, T., Ogura, N., Shimokawa, T., Kakuta, Y.: Fatigue crack growth properties of a GLARE3-5/4 fiber/metal laminate. Engineering Fracture Mechanics **63**(3), 253–272 (1999). doi:10.1016/S0013-7944(99)00021-1
118. Timoshenko, S.: Analysis of Bi-Metal Thermostats. Journal of the Optical Society of America **11**(3), 233 (1925). doi:10.1364/JOSA.11.000233
119. Twigg, G., Poursartip, A., Fernlund, G.: Tool–part interaction in composites processing. Part I: experimental investigation and analytical model. Composites Part A: Applied Science and Manufacturing **35**(1), 121–133 (2004). doi:10.1016/S1359-835X(03)00131-3
120. Van Rooijen, R.G.J.: Bearing strength characteristics of standard and steel reinforced GLARE. Dissertation, TU Delft (2006). URL `https://repository.tudelft.nl/record/uuid:93051a09-e773-49d6-8d7a-bb90474324d8`
121. Verbruggen, M.: Relaxation due to temperature, moisture and external loading — Preliminary results
122. Vidil, T., Tournilhac, F., Musso, S., Robisson, A., Leibler, L.: Control of reactions and network structures of epoxy thermosets. Progress in Polymer Science **62**, 126–179 (2016). doi:10.1016/j.progpolymsci.2016.06.003
123. Vlot, A.: Low-velocity impact loading on fibre reinforced aluminium laminates (ARALL and GLARE) and other aircraft sheet materials. URL `https://repository.tudelft.nl/record/uuid:9200a03f-00ab-4c90-8e38-20748dd5fde0`
124. Vlot, A., Gunnink, J.W.: Fibre metal laminates: An introduction. Springer Netherlands, Dordrecht (2001). URL `https://ebookcentral.proquest.com/lib/gbv/detail.action?docID=3565888`
125. Vlot, A., van Ingen, J.W.: Delamination resistance of post-stretched fibre metal laminates. Journal of Composite Materials **32**(19), 1784–1805 (1998). doi:10.1177/002199839803201903
126. Vooren, A.N., Fagerhus, K.B., Korsvik, K., Joseph, R., Yadav, S., Kumar, A.: Fielding a SHM System for an Aged Military Aircraft. In: S. Farhangdoust, A. Guemes, F.K. Chang (eds.) Structural Health Monitoring 2023. Destech Publications, Stanford, CA (2023)

127. Westre, W.N., Ayers, D.J., Allen-Lilly, H.C., Cregger, S.E., Rothschilds, R.J., Evans, D.W., Hoffman, D.J., Grande, D.L., Rogalski, M.E.: Titanium-polymer hybrid laminates (1997). URL `https://patents.google.com/patent/EP0783960A2/`
128. White, S.R., Hahn, H.T.: Cure cycle optimization for the reduction of processing-induced residual stresses in composite materials. Journal of Composite Materials **27**(14), 1352–1378 (1993). doi:10.1177/002199839302701402
129. Wiedemann, J., Mrzljak, S., Hühne, C., Walther, F.: Increased accuracy of service life prediction for fiber metal laminates by consideration of the manufacturing-induced residual stress state: Proceedings of the 10th European Workshop on Structural Health Monitoring (EWSHM 2024). e-Journal of Nondestructive Testing **29**(7) (2024). doi:10.58286/29832
130. Wisnom, M.R., Gigliotti, M., Ersoy, N., Campbell, M., Potter, K.D.: Mechanisms generating residual stresses and distortion during manufacture of polymer–matrix composite structures. Composites Part A: Applied Science and Manufacturing **37**(4), 522–529 (2006). doi:10.1016/j.compositesa.2005.05.019
131. Withers, P.J., Bhadeshia, H.: Residual stress. Part 1 – Measurement techniques. Materials Science and Technology **17**(4), 355–365 (2001). doi:10.1179/026708301101509980
132. Wu, X.t., Zhan, L.h., Huang, M.h., Zhao, X., Wang, X., Zhao, G.q.: Corrosion damage evolution and mechanical properties of carbon fiber reinforced aluminum laminate. Journal of Central South University **28**(3), 657–668 (2021). doi:10.1007/s11771-021-4635-8
133. Xue, J., Wang, W.X., Takao, Y., Matsubara, T.: Reduction of thermal residual stress in carbon fiber aluminum laminates using a thermal expansion clamp. Composites Part A: Applied Science and Manufacturing **42**(8), 986–992 (2011). doi:10.1016/j.compositesa.2011.04.001
134. Yang, Y., Ng, C.T., Mohabuth, M., Kotousov, A.: Finite element prediction of acoustoelastic effect associated with Lamb wave propagation in pre-stressed plates. Smart Materials and Structures **28**(9), 095007 (2019). doi:10.1088/1361-665X/ab2dd3
135. Yeh, J.R.: Fatigue crack growth in fiber-metal laminates. International Journal of Solids and Structures **32**(14), 2063–2075 (1995). doi:10.1016/0020-7683(94)00221-H
136. Yu, Y., Ashcroft, I.A., Swallowe, G.: An experimental investigation of residual stresses in an epoxy–steel laminate. International Journal of Adhesion and Adhesives **26**(7), 511–519 (2006). doi:10.1016/j.ijadhadh.2005.07.006
137. Zappino, E., Santori, M., Masia, R., Zobeiry, N., Petrolo, M.: Numerical Analysis of the Impact of Process Parameters on the Residual Stress of a Flat Composite Part. Aerotecnica Missili & Spazio (2024). doi:10.1007/s42496-024-00231-7
138. Zobeiry, N., Poursartip, A.: The origins of residual stress and its evaluation in composite materials. In: P.W.R. Beaumont (ed.) Structural integrity and durability of advanced composites, Woodhead publishing series in composites science and engineering, pp. 43–72. Elsevier, Amsterdam (2015). doi:10.1016/B978-0-08-100137-0.00003-1

Chapter 3
Materials and Manufacturing

Abstract The relevant FMLs in this work are manufactured from a stainless steel alloy (X10CrNi18-8) and an epoxy-based CFRP prepreg material (Hexcel 8552-AS4). The relevant elastic and thermal material properties are mainly taken from the literature. For the manufacturing of the FMLs, a standard prepreg process known from the aerospace industry is used, and a mechanical pretreatment step for the steel sheets is added. This additional step includes a vacuum suction blasting process and applying a sol-gel layer. Two fundamentally different cure cycles, the manufacturer-recommended cure cycle (MRCC) and a modified (MOD) cure cycle, allow the generation of significant differences in the residual stress states for all types of laminate configurations. Preliminary quasi-static tensile tests on all four relevant laminate configurations prove the reproducibility of the manufacturing processes and a significant influence of the residual stress states on the ultimate tensile strength and the elastic modulus of the manufactured FMLs.

3.1 Elastic Material Properties

The metal constituent in the fiber metal laminates of this work is a stainless steel alloy (X10CrNi18-8, DIN 1.4310, AISI 301). The elastic material properties of this material are the basis for calculating the residual stresses from measured strains. Due to the rolling process of the steel sheets before FML manufacturing, differences in the tensile elastic modulus between rolling and transverse directions are possible [24]. Therefore, quasi-static tensile tests are carried out on the steel sheets for the two principal in-plane directions. In the course of this work, the rolling direction of the steel sheets is always indicated by the 1-direction (0°-direction) and coincides with the fiber-direction (0°-direction) of the CFRP material. The tensile tests are performed according to DIN EN ISO 6892-1 [7] with rectangular specimens (260 mm × 20 mm). Figure 3.1 shows the mean of all stress-strain data for the rolling direction and the transverse direction and the corresponding standard deviations. Slight variations between the two primary directions are observed.

J. Wiedemann, *Relevance of Manufacturing-Induced Residual Stress States in Fiber Metal Laminates for Structural Health Monitoring*, Mechanics and Adaptronics, https://doi.org/10.1007/978-3-032-09925-9_3

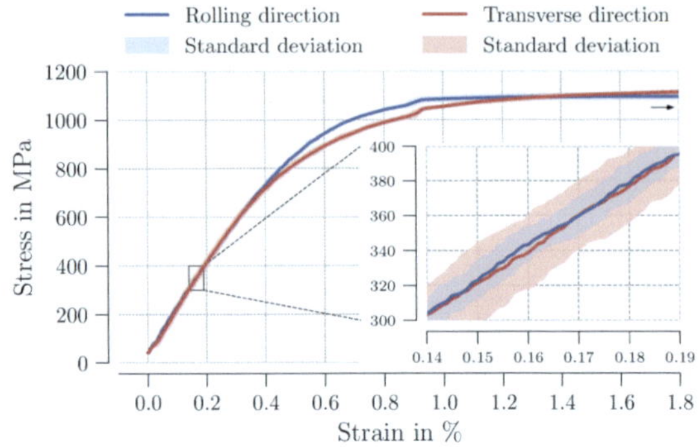

Fig. 3.1 Mean stress-strain data and standard deviation of all steel sheet specimens tested in quasi-static tensile tests in rolling and transverse direction.

Table 3.1: Results of tensile tests on the austenitic steel alloy 1.4310 given as the mean value and the standard deviation. Index 1 indicates the direction of rolling, index 2 denotes the transverse direction.

E_1^t in GPa	E_2^t in GPa	$\sigma_{1,ult}^t$ in MPa	$\sigma_{2,ult}^t$ in MPa
187 ± 2	194 ± 5	1121 ± 7	1147 ± 5

Table 3.2: Elastic material properties (tension) for the austenitic steel alloy 1.4310 used in the following.

E^t in GPa	G in GPa	ν^t	ρ in g cm^{-3}	t_{sheet} in mm
191	73.5[b]	0.3	7.9	0.12 / 0.25

[b] calculated for linear elastic isotropic materials: $G = \frac{E^t}{2(1+\nu^t)}$

Eight specimens for the direction of rolling (E_1^t) and seven specimens for the transverse direction (E_2^t) were evaluated, whereas a consistent behavior can be seen over all the specimens. Consequently, this results in small standard deviations, as shown in Table 3.1. The difference in the tensile elastic modulus for the two principal in-plane directions of around 7 GPa is in accordance with [24].

Since the deviations in the two principal directions are smaller than 5%, isotropic material behavior is assumed for simplicity. All other elastic material properties for the steel used during the subsequent evaluations are taken from the literature and provided in Table 3.2. Steel sheets with a thickness of 0.12 mm are used for most investigations. Only in Chap. 7, some laminates are manufactured with steel sheets having a thickness of 0.25 mm. The same material properties are assumed for the thick steel sheets.

The FRP component used for the FML in this work (Hexcel Hexply 8552-AS4) is a unidirectional carbon-fiber-reinforced prepreg material impregnated with a thermoset (epoxy) resin. The literature reports different values for the elastic material properties. Table 3.3 gives an overview of different literature sources and the reported

Table 3.3: Comparison of elastic material properties for Hexcel Hexply 8552-AS4 from the literature. Values provided by Johnston [16] are used in this work.

Value	Unit	**[16]**	[12]	[22]	[10]	[13]
E_1^t	GPa	**122**	141	132	135	135
$E_2^t = E_3^t$	GPa	**9.9**	10	9.2	9.5	9.5
$G_{12} = G_{13}$	GPa	**5.2**	-	4.8	4.9	4.9
G_{23}	GPa	**3.4**	-	-	*(4.9)*[a]	3.3
$\nu_{12}^t = \nu_{13}^t$	-	**0.27**	-	0.3	0.3	0.3
$\nu_{23}^t = \frac{E_2^t}{2G_{23}} - 1$	-	**0.47**	-	-	0.45	0.45
f. vol. content	%	57.3	57.42	60.38	57.4	-
f. area weight	$\mathrm{g\,m^{-2}}$	-	134	190	-	-
ρ	$\mathrm{g\,cm^{-3}}$	-	1.58	1.57-1.60	-	-
t_{ply} (cured)	mm	-	0.13	0.19	-	0.13

[a] not in accordance with rules of transversal isotropy

values. The most considerable deviations are found in the tensile elastic modulus in the fiber direction E_1^t, which ranges from 122 GPa [16] to 141 GPa [12]. For FRP materials, differences in the elastic material properties can result from manufacturing, testing setup, testing procedure, and even from the evaluation of the raw data. Different evaluation standards (e.g., ISO 527-5, ASTM D 3039, AITM 1-0007) require different ranges in the stress-strain data to calculate the elastic parameters, which can result in significant differences in the elastic modulus.

Contrary to the data reported by [22], the fiber volume content of the prepreg material used in this study, as specified in the Hexcel Corporation data sheet [12], aligns more closely with the values presented by Johnston [16]. Given that Johnston [16] offers the most comprehensive material data set, these values serve as the foundation for the investigations conducted here. Additionally, since the fiber layers in manufactured FMLs are subjected to compression (see Fig. 2.6), it needs to be considered whether compressive elastic properties might provide greater accuracy. As tensile properties are more frequently reported in the literature, they are also utilized in this work. Nonetheless, employing alternative material parameters from the literature or compressive elastic parameters leads to the same conclusions in this work.

3.2 Thermal Material Properties

The CTEs of the materials used in the FMLs are also necessary for the residual stress calculation. Table 3.4 provides the values used in this work. The thermal material properties regarding heat transfer in the involved materials are also necessary for the subsequent chapters. The correlations to derive composite properties from fiber and

Table 3.4 CTEs α of the austenitic steel alloy 1.4310 and the prepreg material 8552-AS4 (Hexcel) in $1 \times 10^{-6}\,\mathrm{K}^{-1}$.

	α_1	α_2
1.4310	19.0	19.0
8552-AS4	0.4	31.2

resin properties and the thermal material characterization for the Hexcel 8552-AS4 prepreg are based on the work of Johnston [16].

According to [16], the specific heat capacity of the composite C_{P_c} from fiber C_{P_f} and resin C_{P_r} properties can be derived using a simple rule of mixture and the fiber volume fraction V_f:

$$C_{P_c} = V_f C_{P_f} + (1 - V_f) C_{P_r} \tag{3.1}$$

However, since the specific heat capacity is a mass-related quantity, it is assumed that an error is made by using the volume fraction, which, however, is small due to the small difference in density between fibers and resin.

Likewise, the thermal conductivity in the fiber direction k_{11_c} of the composite can be derived by the fiber k_{11_f} and resin k_r conductivities:

$$k_{11_c} = V_f k_{11_f} + (1 - V_f) k_r \tag{3.2}$$

For the composite thermal conductivity in the transverse k_{22_c} and thickness direction k_{33_c}, a more comprehensive equation originally proposed by [29] is needed:

$$k_{22_c} = k_{33_c} = k_r \left(\left(1 - 2\sqrt{\frac{V_f}{\pi}} \right) + \frac{1}{B} \left(\pi - \frac{4}{A} \tan^{-1} \frac{A}{1 + B\sqrt{\frac{V_f}{\pi}}} \right) \right) \tag{3.3}$$

$$\text{with} \quad A = \sqrt{1 - \frac{B^2 V_f}{\pi}}, \text{ and} \quad B = 2\left(\frac{k_r}{k_{22_f}} - 1 \right) \tag{3.4}$$

A detailed characterization of the Hexcel 8552-AS4 prepreg material is provided by [16]. The specific heat capacities for the fibers C_{P_f} and for the resin C_{P_r} were derived as:

$$\begin{aligned} C_{P_f} &= 904 + 2.05\,(T - 75)\,/{}^\circ\mathrm{C} \\ C_{P_r} &= 1005 + 3.74\,(T - 20)\,/{}^\circ\mathrm{C} \end{aligned} \tag{3.5}$$

where the temperature T needs to be inserted in °C and the results are given in $\mathrm{J/(kg\,K)}$. The thermal conductivities in fiber direction k_{11_f}, fiber transverse direction k_{22_f} and for the resin k_r are given by:

$$\begin{aligned} k_{11_f} &= 7.69 + 1.56 \times 10^{-2} T/{}^\circ\mathrm{C} \\ k_{22_f} &= 2.4 + 5.07 \times 10^{-3} T/{}^\circ\mathrm{C} \\ k_r &= 0.148 + 3.43 \times 10^{-4} T/{}^\circ\mathrm{C} + 6.07 \times 10^{-2} \alpha \end{aligned} \tag{3.6}$$

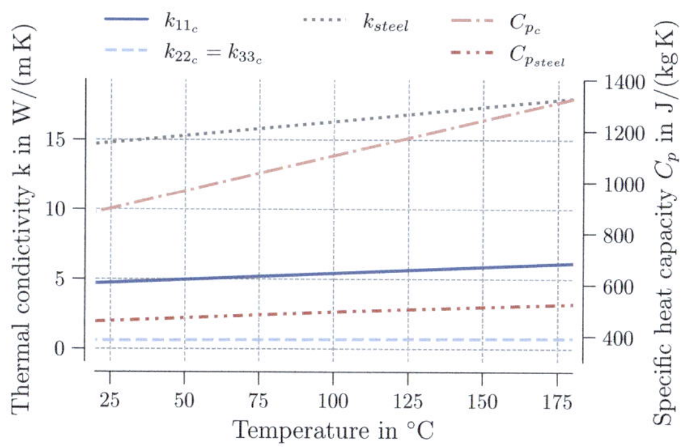

Fig. 3.2 Thermal conductivities and specific heat capacities as a function of temperature for 1.4310 (steel), and as a function of temperature and for a degree of cure of 100% for Hexcel 8552-AS4 (adapted from [35]).

with α being the degree of cure. Again, the temperature is inserted in °C and the results are given in W/(m K).

Consequently, this results in the thermal conductivity k_c and specific heat capacity C_{p_c} for Hexcel 8552-AS4 in the fiber direction (11) and transverse to the fiber direction (22) in dependence of the temperature for a degree of cure of 100% as provided in Fig. 3.2. For the FML steel sheets, the thermal conductivity k_{steel} and specific heat capacity $C_{p_{steel}}$ are also plotted in Fig. 3.2 based on [37].

3.3 Steel Sheet Preparation and Layup

The manufacturing process used for the FMLs in this work is very similar to the manufacturing process for FRP parts using prepreg materials, a standard in the aerospace industry. However, in the production of FMLs, the interface between metal and FRP plays a crucial role. The metal sheets need appropriate pretreatment to ensure good adhesion with the resin and, consequently, a good interlaminar shear strength.

Pretreatment is carried out in a two-step process. The first is a mechanical cleaning step using a vacuum suction blasting process with corundum particles, which is beneficial for thin metal layers where a classical sandblasting process would severely damage the foils [31]. Figure 3.3 shows a photo of the vacuum suction blasting facility used. After the mechanical treatment, the steel sheets are chemically cleaned with heptane to remove dust particles.

The second pre-treatment step involves applying an aqueous sol-gel solution (3M Surface Pre-Treatment AC-130-2). According to the datasheet [1], the metal sheets are dried for one hour after the sol-gel application and laminated immediately with the CFRP prepreg layers.

The mechanical cleaning using a vacuum suction blasting process in combination with an aqueous sol-gel solution has proven to produce mechanically compliant lami-

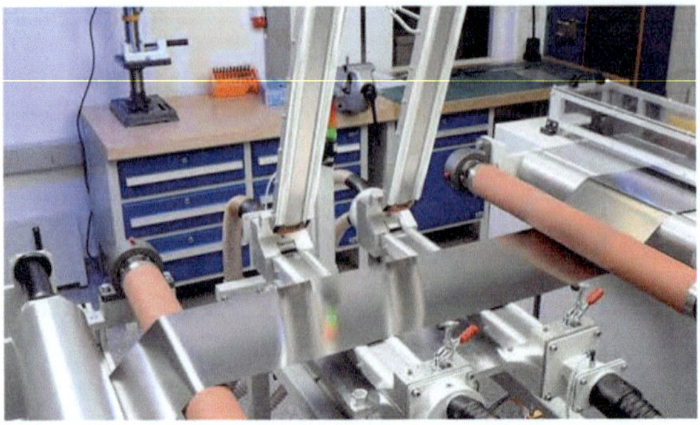

Fig. 3.3 Vacuum suction blasting facility for mechanical pretreatment of the thin steel sheets designed at the German Aerospace Center (DLR) in Braunschweig (adapted from [32]).

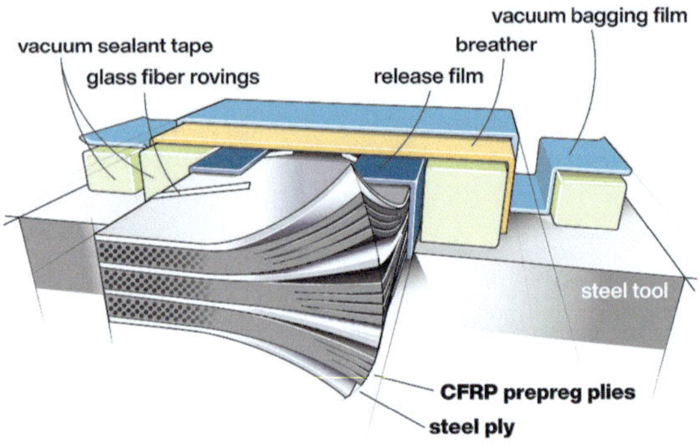

Fig. 3.4 Schematic setup and the use of consumables in the production of the FML test specimens (adapted from [36]).

nates without being environmentally harmful [4,30]. This manufacturing process has also been demonstrated to produce high-quality laminates by other authors [5,27].

The FML laminates are wrapped in a thin release film, with glass fiber rovings placed around the edges to ensure the laminate's degassing. Subsequently, the uncured laminate is placed on a tool, and vacuum sealant tape is used as a border around it. This is illustrated in Fig. 3.4. The FML layups are subsequently cured in an autoclave, except for the laminates investigated in Chap. 4, which are cured in a laboratory oven to reduce the complexity of the experimental setup. Compared to autoclave curing, no additional hydrostatic pressure is used during oven cure. The cover plate used on top of the laminates during autoclave cure is also eliminated in the oven processes.

To assess the homogeneity of the prepreg consolidation and to illustrate the insignificant thickness of the sol-gel layer applied to the steel sheets, laminate cross-sections are prepared and investigated using light microscopy (Keyence VHX-5000) and scanning electron microscopy (Phenom XL). Figure 3.5 shows the results and proves a good laminate homogeneity and the insignificant thickness of the sol-gel layer. Other metal pre-treatment methods produce significantly greater interface

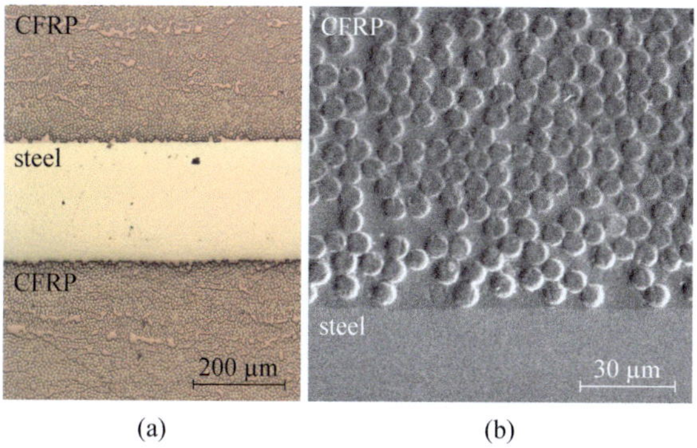

Fig. 3.5 The homogeneity of the consolidated prepreg and the insignificant thickness of the sol-gel layer between CFRP and steel is illustrated in microscopic cross sections of the manufactured laminates: **(a)** light microscopy and **(b)** scanning electron microscopy (adapted from [34]).

Table 3.5: Laminate configurations used in this work.

ID	Symbol	Layup	t_{steel} [mm]	t_{lam} [mm]	MVF [%]	NoI [-]	FTF [%]	Chap.
2	○	$[St/0°_4/St/0°_2]_S$	0.12	2.04	23.5	6	11.8	4, 6, 7
3a	□	$[(St/0°_4)_5/\overline{St}]_S$	0.12	6.52	20.2	20	3.7	7
4	△	$[(St/0°)_2/St/\overline{0°}]_S$	0.25	2.15	69.8	10	23.3	6, 7
6b	◇	$[(St/0°_4)_3/\overline{St}]_S$	0.25	4.87	35.9	12	10.3	7

t: thickness, MVF: metal volume fraction, NoI: number of interfaces,
FTF: facesheet thickness fraction

thicknesses between the fiber and metal layers. The importance of the interface layer thickness in material fatigue was shown in [3], since it governs delamination growth.

In this work, different FML layup configurations are manufactured and investigated in various experimental setups. Table 3.5 presents the most important laminate configurations. All laminates have a constant metal volume fraction and laminate thickness. The facesheets (outer layers) of all layups are steel. The layups differ by their respective metal volume fraction, Eq. (2.1), and the number of interfaces (NoI) between steel and CFRP:

$$\mathrm{NoI} = 2\,N_{\mathrm{steel\ sheets}} - 2 \tag{3.7}$$

with $N_{\mathrm{steel\ sheets}}$ being the total number of steel sheets in the laminate. Furthermore, they also differ by their face sheet thickness fraction (FTF):

$$\mathrm{FTF} = \frac{2\,t_{\mathrm{steel\ sheet}}}{t_{\mathrm{laminate}}} \tag{3.8}$$

The configuration labeled with ID2 is the layup used during all investigations to ensure comparability. With an MVF of 23.5%, this configuration contains a significant amount of residual stresses as is indicated in Fig. 2.6. Table 3.5 also specifies the chapters in which the various laminate configurations are used.

The laminates needed in Chap. 4 and 6 are manufactured with the geometrical dimensions as needed, while the laminates in Chap. 5 and 7 are cut to dimension

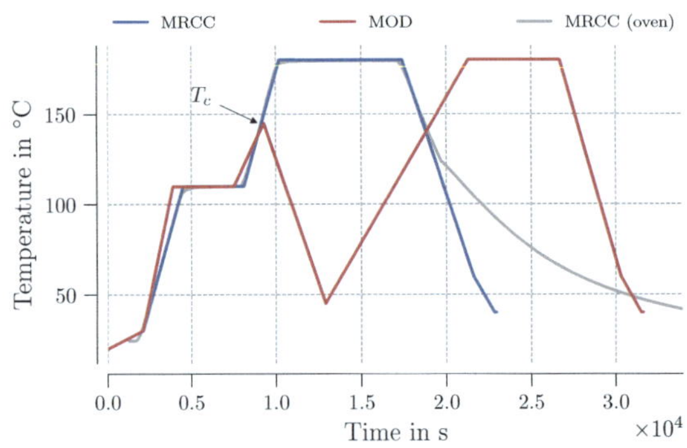

Fig. 3.6 Temperature over time for the manufacturer-recommended cure cycle (MRCC) and the modified cure cycle (MOD) used in this work. The oven cycles deviate from the autoclave cycles during cooling since the lab oven does not provide an active cooling function.

using water jet cutting. With this cutting process, no heat is introduced into the specimen, and edge delaminations, as observed with saw cutting, are avoided. The holes of the open hole tension specimens in Chap. 7 are drilled using a 5 mm drill bit.

3.4 Cure Cycles

Two fundamentally different cure cycles are used in this work to generate different residual stress states in the laminates. One is the manufacturer-recommended cure cycle (MRCC), and the other is a modified (MOD) cure cycle. Figure 3.6 shows the temperatures over time for the two main types of cure cycles. While the MRCC is based on the prepreg datasheet [12], the modified cure cycle for the Hexcel 8552-AS4 resin was developed by Prussak et al. [28]. The modifications start from the end of the dwelling stage, where the temperature gradient is reduced. After reaching the cooling temperature T_c, a cooling step is initiated. The goal is to start the curing reaction in the resin before cooling the laminate. The chemical curing reaction responds to a temperature change with a certain time lag and is not completely stopped by a temperature drop once initiated. Thus, it can be achieved that the laminate is at a lower temperature when passing the gel point or showing a substantial amount of elastic behavior. After the intermediate cooling step in the MOD cycle, the laminate is slowly heated to the same maximum curing temperature as in the MRCC to ensure the same final degree of cure.

In Chap. 4, slight variations of the MOD cycle that differ in the cooling temperature and the heating and cooling rates are used to show several cure cycle influences on the curing strains. These MOD variations are introduced at the relevant places in this work. Slight differences are also observed in the cooling stages of the cure cycles when comparing autoclave cure and lab oven cure. Since the lab oven does not possess an active cooling system, the temperature during cooling will not decrease linearly as seen with autoclave cure.

Table 3.6: Process steps of the autoclave manufacturing processes MRCC and MOD.

	Step	1	2	3	4	5	6	7	8	9	10
MRCC	Time in min	0	35	40	60	35	120	70	20	-	-
	Temperature in °C	20	30	110	110	180	180	60	40	-	-
	Pressure in bar	0	7	7	7	7	7	7	0	-	-
	Vacuum in bar	-1	-1	-1	-1	-1	-1	-1	0	-	-
MOD	Time in min	0	35	30	60	30	60	140	90	70	20
	Temperature in °C	20	30	110	110	145	45	180	180	60	40
	Pressure in bar	0	7	7	7	7	7	7	7	7	0
	Vacuum in bar	-1	-1	-1	-1	-1	-1	-1	-1	-1	0

Table 3.6 shows the cure cycles in terms of time steps which can be directly used for autoclave or oven control.

3.5 Resin Cure Kinetics

The literature provides several cure kinetics models to determine the degree of cure during the manufacturing process. Generally, the basic structure of the cure kinetics models proposed in the literature is a first-order differential equation [19, 25]:

$$\frac{\partial \alpha}{\partial t} = K(T)(1-\alpha)^n \tag{3.9}$$

with the degree of cure α, the time t and a material dependent parameter n. The temperature dependence of the reaction is often modeled with an Arrhenius relation:

$$K(T) = A\,e^{\frac{-\Delta E}{RT}} \tag{3.10}$$

with the pre-exponantial factor A which is a material dependent constant, the activation energy ΔE, the gas constant R and the instantaneous resin temperature T.

Other models include the autocatalytic behavior of the curing reaction, which better represents the sigmoidal behavior of the curing reaction known from experimental analyses [17, 18]:

$$\frac{\partial \alpha}{\partial t} = K(T)\,\alpha^m\,(1-\alpha)^n \tag{3.11}$$

where m again is a material-dependent constant. These models, however, need a starting value for $\alpha > 0$ or need to be combined with the model from Eq. (3.9).

More advanced models consider that after vitrification, the movement of the molecules is increasingly restricted, and the curing reaction is significantly slowed down and becomes diffusion controlled [6, 15]:

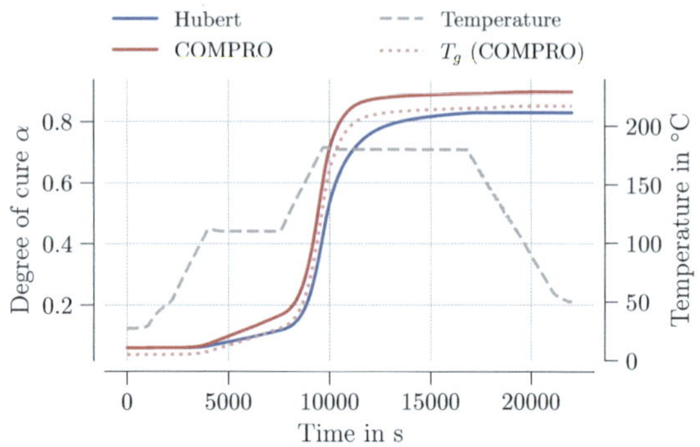

Fig. 3.7 Comparison of two cure kinetics models for the Hexcel 8552-AS4 resin and the manufacturer-recommended cure cycle. The given glass transition temperature (T_g) is based on the degree of cure of the COMPRO model.

$$\frac{\partial \alpha}{\partial t} = \frac{K(T)\,\alpha^m\,(1-\alpha)^n}{1 + e^{C(\alpha - (\alpha_{C0} + \alpha_{CT}\,T))}} \tag{3.12}$$

The model parameters for the Hexcel 8552-AS4 resin for the cure kinetics model given in Eq. (3.12) are provided in Hubert et al. [15]. A more comprehensive material characterization of the Hexcel 8552-AS4 resin and a corresponding cure kinetics model is part of the commercial software COMPRO of Convergent Manufacturing Technologies, Inc. The mathematical formulation of the model and the respective parameters for the Hexcel 8552-AS4 resin are provided in [33]. Figure 3.7 compares the degree of cure of the Hexcel 8552-AS4 resin over time for the temperature data of the MRCC and the two respective cure kinetics models. The glass transition temperature based on the degree of cure of the COMPRO model is also provided in the figure. The instantaneous glass transition temperature is calculated using the Di-Benedetto correlation [23, 33]:

$$T_g = T_{g0} + \frac{\lambda\alpha(T_{g\infty} - T_{g0})}{1 - (1-\lambda)\alpha} \tag{3.13}$$

with λ being a material-specific parameter derived from the initial and final specific heat capacities before and after cure, and T_{g0} and $T_{g\infty}$ being the initial and final glass transition temperatures.

As discussed in Sec. 2.2.1, from gelation onwards, the resin possesses significant stiffness to transfer loads. Therefore, the gel-point is particularly interesting to understand residual stress formation. Based on statistical considerations of Flory [9] the gel-point can be calculated using the functionalities f of the resin components when assuming that resin and curing agent are mixed in stoichiometric ratio:

$$\alpha_{\text{gel}} = \frac{1}{\sqrt{(f_{\text{resin}} - 1)(f_{\text{curing agent}} - 1)}} \tag{3.14}$$

The single chemical components of the Hexcel 8552-AS4 resin are provided in the safety data sheet [11] and given in Table 3.7. However, the share of the single

Table 3.7: Chemical components of the Hexcel 8552-AS4 epoxy resin and their functionalities [11].

Component	Percentage	Functionality
4,4'-Methylenebis[N,N-bis(2,3-epoxypropyl)aniline]	15% to 40%	4
p-(2,3-epoxypropoxy)-N,N-bis(2,3-epoxypropyl)aniline	15% to 40%	3
3,3'-sulphonyldianiline	10% to 30%	4
DAPSON	1% to 15%	4

components is only given as a range. Since the resin consists of two components with functionalities 3 and 4, the gel point, according to Flory, can only be estimated to be somewhere below $\alpha_{gel} = 0.41$. In [14], the gel point of the Hexcel 8552-AS4 resin was derived in rheological experiments to $\alpha_{gel} = 0.469$ and also used in [16]. Whereas the definition of the gel point in rheological experiments is defined by either the correlation of storage and loss modulus or by a viscosity limit, which is substantially different from the statistical considerations of Flory [9].

The gel point indicates where temperature control is crucial during the cure cycle, particularly for MOD cure cycles, to achieve a desired residual stress state. Once gelation occurs, the resin transitions from a viscous to a visco-elastic state and eventually to an elastic state upon vitrification. After gelation, the resin can transfer loads, which is essential for the residual stress evolution. The final residual stress state at room temperature depends on the temperature at which the initially independent laminate layers start to interact. This happens after the transition of the resin into the visco-elastic state.

3.6 Preliminary Investigations - Laminate Strength and Cure Cycle Influence

To validate the manufacturing of the CFRP-steel laminates and the effectiveness of the modified cure cycle, quasi-static tensile tests[1] are performed on two FML specimens for each laminate configuration from Table 3.5 and cure cycle. Based on ISO 527-5 [8] rectangular specimens (type A) are manufactured. Since the specimens are cut out of a plate manufactured for specimens used in Chap. 7, the size of the specimens is 180 mm x 20 mm and therefore, the gauge length is reduced to 80 mm.

For the tensile tests, an Instron 8802 (Instron, High Wycombe, UK) servo-hydraulic testing system with a load capacity of 250 kN is used in combination with an Instron dynamic extensometer (CAT No. 2620-603 with 10 mm gauge length, ±1 mm travel) to record strain. The testing speed is set to 1 mm min^{-1}. The specimens with ID3a (thickness of 6.52 mm) could not be tested until final failure due to the limited load capacity of the tensile testing system.

[1] The experiments were performed in cooperation with Selim Mrzljak from the Chair of Materials Test Engineering (WPT) of the TU Dortmund University.

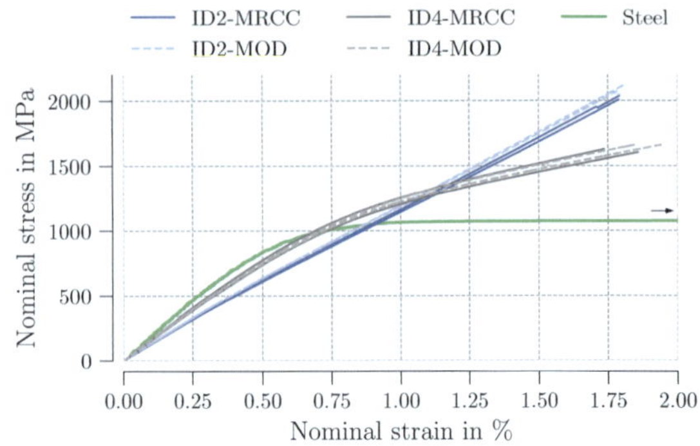

Fig. 3.8 Quasi-static nominal tensile stress vs. strain for the laminate configurations ID2 and ID4 and results of the quasi-static tensile tests on the steel sheet (rolling direction) as a reference (adapted from [34]).

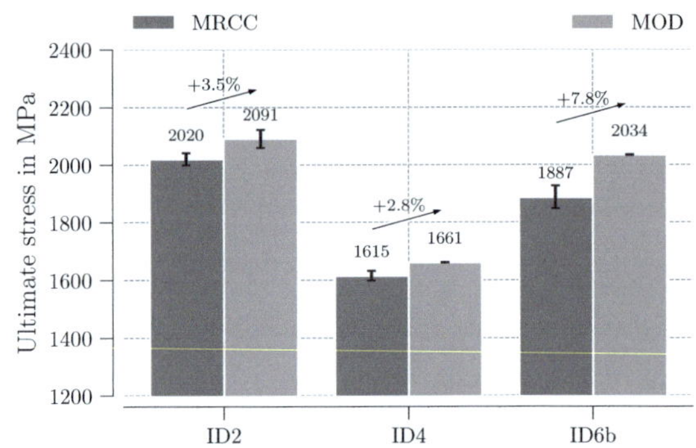

Fig. 3.9 The comparison of the ultimate tensile strength for laminate configurations ID2, ID4, and ID6b indicates that using the MOD cure cycle improves the ultimate laminate strength.

Figure 3.8 shows the stress-strain data for specimens ID2 (MVF: 23.5%) and ID4 (MVF: 69.8%). Additionally, the stress-strain data of the steel sheet (rolling direction) is added as a reference (cf. Fig. 3.1). It is visible that with increasing MVF, the stress-strain curve better resembles the ductile behavior of the pure steel specimen (ID2 vs ID4). For lower MVF, the typical almost linear stress-strain behavior of CFRP is more evident in the data. The final failure of the FMLs occurs at approximately the same total strain for both layups, which is typical for FMLs in general, as the failure strain of the fibers governs ultimate laminate failure [2].

The tensile test results further show that the modified cure cycle results in higher ultimate tensile strengths of the specimens independent of the MVF (Fig. 3.9). A significant increase in the ultimate tensile strength is found for all specimens manufactured with the modified cure cycle. This is in good agreement with the literature [20, 21, 26].

A similar picture can be found when evaluating the elastic modulus of the FML test specimens. The modulus evaluation is done in the strain range of 0.05% to 0.25% according to ISO 527-5 [8]. Table 3.8 shows that an increase in modulus is found for the specimens where the residual stresses are assumed to be significantly reduced using the modified cure cycle (ID2, ID3a, and ID6b). For the specimens with

Table 3.8: Evaluation of the elastic modulus in the strain interval 0.05% to 0.25% (according to DIN EN ISO 527-5 [8]).

	ID2	ID3a	ID4	ID6b	
MRCC	127 ± 2	130 ± 5	159 ± 8	137	GPa
MOD	131	136 ± 3	157 ± 2	138 ± 2	GPa
Delta	+3.0	+4.6	-1.3	+1.1	%

layup ID4, the influence of the cure cycle is expected to be small since the MVF is relatively high (69.8%) and, consequently, the residual stresses are relatively small, as indicated in Fig. 2.6.

The results of the quasi-static tensile tests prove that the manufacturing processes used in this work allow the production of CFRP-steel laminates with high mechanical quality and reproducible properties (minor deviations between identical tested specimens). Furthermore, the results show that using the two different cure cycles, MRCC and MOD, a significant difference in the mechanical properties of the FML is achieved, which can be attributed to the different residual stress states in the laminates. Therefore, the identified processes, cure cycles and laminate configurations are a valid parameter set to investigate the residual stress evolution and its influence on other SHM-related quantities in the next chapters.

References

1. 3M Aerospace and Commercial Transportation Division: 3M™ surface pre-treatment AC-130-2 (2015). URL `https://multimedia.3m.com/mws/media/10638340/3m-surface-pre-treatment-ac-130-2-application-guide.pdf`
2. Alderliesten, R.: Fatigue and fracture of fibre metal laminates, *Solid mechanics and its applications*, vol. 236. Springer, Cham (2017)
3. Alderliesten, R.C.: On crack tunneling and plane-strain delamination in laminates. International Journal of Fracture **148**(4), 401–414 (2007). doi:10.1007/s10704-008-9212-8
4. Blohowiak, K.Y., Osborne, J.H., Krienke, K.A.: Surface pretreatment of metals to activate the surface for sol-gel coating (1999). URL `https://patents.google.com/patent/US5869140A/en`
5. Boose, Y., Kappel, E., Stefaniak, D., Prussak, R., Pototzky, A., Weiß, L.: Phenomenological investigation on crash characteristics of thin layered CFRP-steel laminates. International Journal of Crashworthiness **182**(15), 1–10 (2020). doi:10.1080/13588265.2020.1787681
6. Cole, K.C., Hechler, J.J., Noel, D.: A new approach to modeling the cure kinetics of epoxy/amine thermosetting resins. 2. Application to a typical system based on bis[4-(diglycidylamino)phenyl]methane and bis(4-aminophenyl) sulfone. Macromolecules **24**(11), 3098–3110 (1991). doi:10.1021/ma00011a012
7. DIN Deutsches Institut für Normung e.V.: DIN EN ISO 6892-1:2020-06, Metallische Werkstoffe-Zugversuch-Teil 1: Prüfverfahren bei Raumtemperatur (ISO 6892-1:2019); Deutsche Fassung EN_ISO_6892-1:2019. doi:10.31030/3132591
8. DIN Deutsches Institut für Normung e.V.: Plastics - Determination of tensile properties - Part 5: Test conditions for unidirectional fibre-reinforced plastic composites (ISO 527-5:2021) (2022)

9. Flory, P.J.: Molecular Size Distribution in Three Dimensional Polymers. I. Gelation 1. Journal of the American Chemical Society **63**(11), 3083–3090 (1941). doi:10.1021/ja01856a061
10. Garstka, T.: Separation of process induced distortions in curved composite laminates. Dissertation, The University of Bristol, Bristol (2005)
11. Hexcel Corporation: Safety data sheet HexPly 8552 (2015). URL `https://www.hexcel.com/user_area/content_media/raw/HexPly_8552_eu_DataSheet.pdf`
12. Hexcel Corporation: Product data sheet HexPly 8552 (2020). URL `https://www.hexcel.com/user_area/content_media/raw/HexPly_8552_eu_DataSheet.pdf`
13. Hörberg, E., Nyman, T., Åkermo, M., Hallström, S.: Thickness effect on spring-in of prepreg composite L-profiles – An experimental study. Composite Structures **209**(8), 499–507 (2019). doi:10.1016/j.compstruct.2018.10.090
14. Hubert, P.: Aspects of flow and compaction of laminated composite shapes during cure. Dissertation, The University of British Columbia, Vancouver, British Columbia, Canada (1996). doi:10.14288/1.0078499
15. Hubert, P., Johnston, A., Poursartip, A., Nelson, K.: Cure Kinetics and Viscosity Models for Hexcel 8552 Epoxy Resin. In: L. Repecka, F.F. Saremi (eds.) A materials and processes odyssey, Science of advanced materials and process engineering series. Society for the Advancement of Material and Process Engineering, Covina, California (2001)
16. Johnston, A.A.: An integrated model of the development of process-induced deformation in autoclave processing of composite structures. Dissertation, University of British Columbia, Vancouver, Kanada (1997). doi:10.14288/1.0088805
17. Kamal, M.R., Sourour, S.: Kinetics and thermal characterization of thermoset cure. Polymer Engineering and Science **13**(1), 59–64 (1973). doi:10.1002/pen.760130110
18. Keenan, M.R.: Autocatalytic cure kinetics from DSC measurements: Zero initial cure rate. Journal of Applied Polymer Science **33**(5), 1725–1734 (1987). doi:10.1002/app.1987.070330525
19. Kenny, J.M., Trivisano, A.: Isothermal and dynamic reaction kinetics of high performance epoxy matrices. Polymer Engineering and Science **31**(19), 1426–1433 (1991). doi:10.1002/pen.760311909
20. Kim, H.S., Park, S.W., Lee, D.G.: Smart cure cycle with cooling and reheating for co-cure bonded steel/carbon epoxy composite hybrid structures for reducing thermal residual stress. Composites Part A: Applied Science and Manufacturing **37**(10), 1708–1721 (2006). doi:10.1016/j.compositesa.2005.09.015
21. Kim, H.S., Yoo, S.H., Chang, S.H.: In situ monitoring of the strain evolution and curing reaction of composite laminates to reduce the thermal residual stress using FBG sensor and dielectrometry. Composites Part B: Engineering **44**(1), 446–452 (2013). doi:10.1016/j.compositesb.2012.04.021
22. National Institute for Aviation Research - Wichita State University: Hexcel 8552 AS4 unidirectional prepreg at 190 gsm & 35% RC: Qualification material property data report (2011). URL `https://www.wichita.edu/research/NIAR/Research/hexcel-8552/AS4-Unitape-2.pdf`
23. Pascault, J.P., Williams, R.J.J.: Relationships between glass transition temperature and conversion. Polymer Bulletin **24**(1), 115–121 (1990). doi:10.1007/BF00298330
24. Petersen, E., Stefaniak, D., Hühne, C.: Experimental investigation of load carrying mechanisms and failure phenomena in the transition zone of locally metal reinforced joining areas. Composite Structures **182**(6), 79–90 (2017). doi:10.1016/j.compstruct.2017.09.002
25. Prime, R.B.: Differential scanning calorimetry of the epoxy cure reaction. Polymer Engineering and Science **13**(5), 365–371 (1973). doi:10.1002/pen.760130508
26. Prussak, R., Stefaniak, D., Hühne, C., Sinapius, M.: Residual stresses in intrinsic UD-CFRP-steel-laminates - Experimental determination, identification of sources, effects and modification approaches. In: 20th Symposium on Composites, Materials Science Forum, pp. 369–376. Trans Tech Pubn (2015). doi:10.4028/www.scientific.net/MSF.825-826.369. URL `https://www.scientific.net/MSF.825-826.369.pdf`
27. Prussak, R., Stefaniak, D., Hühne, C., Sinapius, M.: Evaluation of residual stress development in FRP-metal hybrids using fiber Bragg grating sensors. Production Engineering **12**(2), 259–267 (2018). doi:10.1007/s11740-018-0793-4

28. Prussak, R., Stefaniak, D., Kappel, E., Hühne, C., Sinapius, M.: Smart cure cycles for fiber metal laminates using embedded fiber Bragg grating sensors. Composite Structures **213**, 252–260 (2019). doi:10.1016/j.compstruct.2019.01.079
29. Springer, G.S., Tsai, S.W.: Thermal conductivities of unidirectional materials. Journal of Composite Materials **1**(2), 166–173 (1967). doi:10.1177/002199836700100206
30. Stefaniak, D.: Improving residual strength of unidirectionally reinforced plastic laminates by metal layering. Dissertation, Technische Universität Carola-Wilhelmina zu Braunschweig (2017)
31. Stefaniak, D., Kappel, E., Kolesnikov, B., Hühne, C.: Improving the mechanical performance of unidirectional CFRP by metal-hybridization. In: ECCM15 - 15th European Conference on Composite Materials (2012)
32. Stefaniak, D., Prussak, R., Weiß, L.: Spezifische Herausforderungen für den Einsatz von Faser-Metall-Laminaten. Lightweight Design **10**(5), 24–31 (2017). doi:10.1007/s35725-017-0046-z
33. Van Ee, D., Poursartip, A.: HexPly 8552 material properties database for use with COMPRO CCA and Raven (2009)
34. Wiedemann, J., Mrzljak, S., Koord, J., Hühne, C., Walther, F.: Modified cure cycles for increased fatigue performance of fiber metal laminates. Composite Structures **351**, 118631 (2025). doi:10.1016/j.compstruct.2024.118631
35. Wiedemann, J., Pittsyk, T., Völkerink, O., Hühne, C.: Temperature distribution inside composite and fiber metal laminates during modified cure cycles. Procedia CIRP **131**, 107–112 (2025). doi:10.1016/j.procir.2024.10.333
36. Wiedemann, J., Prussak, R., Kappel, E., Hühne, C.: In-situ quantification of manufacturing-induced strains in fiber metal laminates with strain gages. Composite Structures **Vol. 691**(1), 115967 (2022). doi:10.1016/j.compstruct.2022.115967
37. Zapp Precision Metals GmbH: Rostfreier Federbandstahl 1.4310: Kennwerte für 1.4310. URL `https://www.zapp.com/fileadmin/_documents/Downloads/PS/informationsdatenblaetter/Federbandkatalog_PS.pdf`

Chapter 4
Residual Stress Quantification Using Strain Gauges

Abstract Strain gauges attached to the metal sheets in an FML can measure the strains that develop during the curing process from the beginning on. This is an advantage compared to measurements in FRP, where the connection between the sensor and substrate is only established during the curing process. The measured strains across various cure cycles show that significant residual stresses evolve in the metal sheets of a CFRP-steel laminate with a metal volume fraction of 23.5%. With modified cure cycles, these residual stresses can be reduced by up to 50%. The strain signal over the curing process further indicates the point in time the CFRP resin possesses a significant shear modulus and can consequently transfer loads between the metal and fibers, which leads to an interaction between the two FML components. The strain gauge measurements are validated with integrated fiber Bragg grating sensors. Furthermore, temperature measurements and the development of a numerical heat transfer model prove the validity of the measurements at only discrete points in the laminate, and the transfer of these results into a residual stress state of the entire laminate. The results further show that assuming the bonding temperature equals the stress-free temperature is only valid for bonding temperatures close to the final curing temperature. When the connection is established at lower temperatures, significantly below the final curing temperature, errors are made when calculating the residual stress state using this assumption.[1]

4.1 Motivation

A comprehensive quantification of the residual stress state is necessary to correlate the influence of the residual stress state in FMLs with the SHM tasks of interest. Furthermore, investigating the development of the residual stress state during the curing process in detail for all relevant cure cycles helps to understand the influence

[1] Parts of the results of this chapter have been published in [32], [31], and [10]. Minor deviations of some parameters from earlier publications result from differences in the material properties assumed in these studies.

J. Wiedemann, *Relevance of Manufacturing-Induced Residual Stress States in Fiber Metal Laminates for Structural Health Monitoring*, Mechanics and Adaptronics, https://doi.org/10.1007/978-3-032-09925-9_4

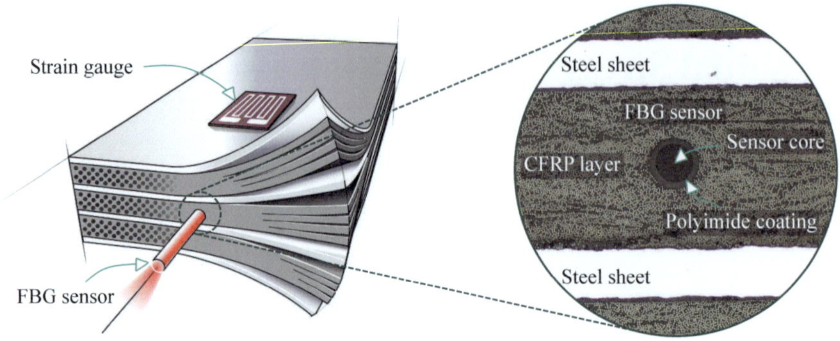

Fig. 4.1: Fiber metal laminate instrumented with a strain gauge and a fiber Bragg grating sensor, and microsection of the FBG sensor embedded between two CFRP plies in the CFRP-steel laminate.

of cure cycle modifications on the residual stress built up. This will validate the modified cure cycles and further prove their applicability for the investigations in this work.

There are several experimental methods to quantify the process-induced residual stress state in composite materials. Zobeiry [36] gives a good overview of common techniques and relevant literature sources. The methods can be distinguished by the time of application with respect to the manufacturing process. During manufacturing, mainly embedded sensors (e.g., fiber Bragg gratings (FBGs) or strain gauges) are used, whereas, after manufacturing, there are several non-destructive (e.g., curvature measurement) and destructive methods (e.g., layer removal). All the different methods have drawbacks regarding temperature sensitivity, bonding issues, accuracy, and others.

This chapter focuses on methods that can provide information during the manufacturing process to understand the development of residual strains and stresses during cure. The destructive and non-destructive methods only provide a final residual stress state. In contrast, using embedded or applied sensors provides time-based data throughout the entire manufacturing cycle. A disadvantage of sensors integrated into the manufacturing setup is their high sensitivity to temperature [36]. However, with accurate temperature measurement and appropriate compensation methods, this disadvantage can be compensated [10, 22].

The most common techniques for measuring strains evolving during the manufacturing process in composite materials are FBG sensors [2, 9, 10, 14, 19, 22, 27] or strain gauges (SGs) [4–6, 11, 29, 33]. Both sensor types can also be easily used for measuring the curing strains during manufacturing in FMLs, as illustrated in Fig. 4.1.

FBG sensors consist of a single glass fiber with a protective coating. They can be embedded directly into the FRP without disturbing the laminate on a macro level when oriented in the fiber direction, as seen from the microsection in Fig. 4.1. During the curing of the resin, the FBG sensor binds to the matrix and, after binding, experiences strains in the same way as the material around it. However, because the binding only develops during cure, it is impossible to measure strains in a laminate from the beginning of the manufacturing process [23].

Likewise, strain gauges are placed inside the laminate before cure, but the same problem arises as with the FBG sensors in terms of absolute measurements. Therefore, Wisnom et al. [33] used a spot curing technique to measure strains from the beginning of the manufacturing process. However, an embedded strain gauge represents a more significant discontinuity than an FBG sensor. Twigg, Poursartip, and Fernlund [29] used SGs on the tool surface to monitor the interaction between a laminate and the metal tool. However, this only provides information about the stress transfer between the laminate and the tool surface and not the residual stress state in the laminate itself.

In the context of FMLs, there is one great advantage compared to monolithic FRP regarding residual stress quantification. The metal constituent in an FML behaves purely linear elastic during the entire manufacturing process. Thus, it is assumed that the strains in the metal sheets can be quantified over the total time of the process. This provides an advantage over methods where the load-carrying bond between the sensor and base material is only established during the manufacturing process, and recorded strains can only be interpreted from this point onwards. This characteristic of FMLs will help to understand the strain evolution during modified cure and to validate these cure cycles. To investigate this, the following hypothesis is formulated:

! Research hypothesis 1

In contrast to FRPs, strain sensors attached to the metal layers of FMLs enable the recording of meaningful time-based strain data from the beginning of the manufacturing process.

Since strain gauges are inexpensive and easy to apply to metal surfaces, the focus is on curing strain measurements using strain gauges on the metal sheets of CFRP-steel laminates. Additionally, FBG sensors are used to validate the strain gauge signals.

4.2 Strain Gauge Technique

In strain gauge measurements[2], the proportional relation between a strain and the change in resistance of a sensor is utilized. The sensor-specific gauge factor F_G correlates the change in resistance $\Delta R / R_0$ with the strain ε:

$$\frac{\Delta R}{R_0} = F_G \, \varepsilon \tag{4.1}$$

[2] The development of the strain gauge measurement method is based on the author's previous work: Wiedemann, J.: A study on process-induced residual stresses and distortions of GLARE structures. Master's thesis, Technische Universität Braunschweig (2018)

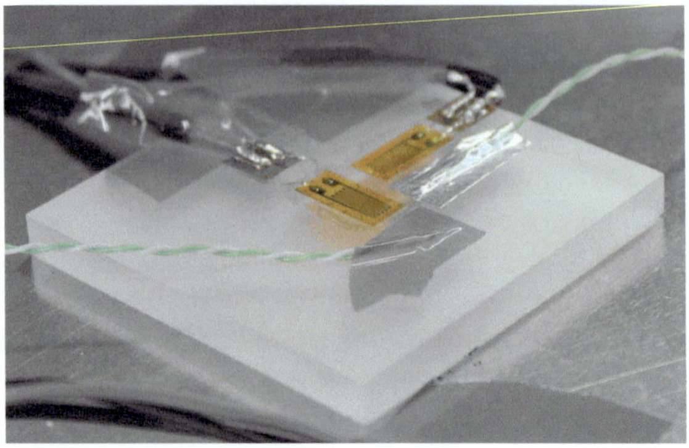

Fig. 4.2 Two dummy strain gauges and two thermocouples attached to an ultra-low expansion glass specimen, which serve as a reference to compensate for temperature-induced strains (adapted from [10]).

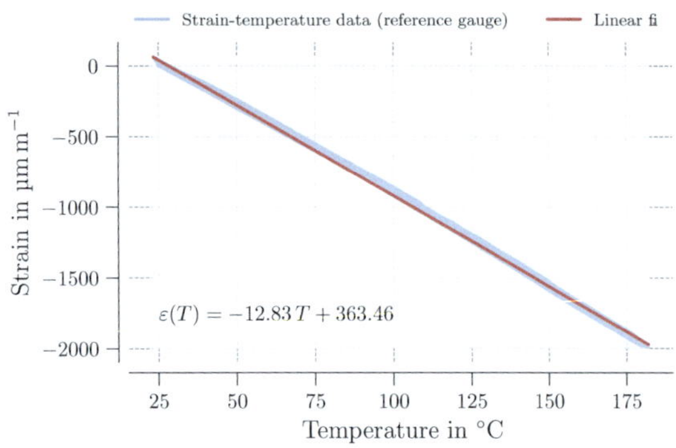

Fig. 4.3 Compensation master curve to compensate for temperature-induced strains when measuring with 120 Ω strain gauges (HBM LC11-6/120) for a starting temperature of 25 °C.

However, this relation is no longer valid when measuring under variable temperature [15, 18, 28]. A reference strain gauge attached to a silicate glass specimen accounts for the influence of temperature. The glass specimen (ultra-low expansion (ULE) glass) possesses a CTE of $0 \pm 0.03 \times 10^{-6}\,\mathrm{K}^{-1}$. The glass specimen is placed within the cure cycle on top of the vacuum bag. Consequently, the reference gauge only measures temperature-induced effects unrelated to thermal or curing strains in the material of interest. By subtracting the measured strains of the reference strain gauge from the measured signal of the strain gauge attached to the FML, all distracting effects due to the temperature change can be eliminated [28]. Figure 4.2 shows a picture of two reference strain gauges attached to the ULE glass specimen.

For the experiments, 120 Ω strain gauges (HBM LC11-6/120) are set up in a three-wire hook-up of a Wheatstone-bridge (quarter-bridge arrangement) to further compensate for any disturbances in the wires themselves.

From the six relevant measurements performed in this work, all the strain-temperature data acquired with the reference strain gauge on the ULE glass specimen is plotted into a single graph in Fig. 4.3. Subsequently, a linear fit is calculated for all the data points. With this compensation master curve, a dummy strain gauge is no longer necessary for measurements using the same setup as in this work. The following equation directly gives the temperature-induced strain $\varepsilon(T)$, while T_0 is the starting temperature of the experiment:

$$\varepsilon(T) = -12.83T + 363.46 + (T_0 - 25) \tag{4.2}$$

Independent of the temperature compensation, the final residual strains at the end of each cure cycle do not need to be corrected for any temperature influence when each cure cycle's start and final temperatures are the same. From the recorded in-plane residual strains ε_{res} the resulting residual stresses σ_{res} in the metal layer can be calculated by applying Hooke's law using the reduced stiffness matrix $[Q]_k$ for the respective ply:

$$\{\sigma_{res}\}_k = [Q]_k \{\varepsilon_{res}\}_k \tag{4.3}$$

For all the measurements in this chapter, square laminates with an edge length of 125 mm and the layup ID2 (cf. Table 3.5) are manufactured. The strain gauges are positioned in the center of the top metal layer. For the combined experiment with strain gauges and an FBG sensor, the FBG is placed between the first and second CFRP ply with the Bragg grating positioned in the center of the laminate. The thermocouple to measure the laminate temperature is always placed between the first and second CFRP plies but only 10 mm from the edge to not disturb the strain measurements.

4.2.1 Evaluation of Measured Strains During Cure

This section examines the recorded strains on the top metal sheets of the FMLs produced with various cure cycles. The curing strains are analyzed over the process time and the process temperature. The latter allows a better correlation of the recorded strains to the polymer cure kinetics.

Curing Strains vs Process Time

Figure 4.4 shows the strains recorded during the MRCC over time for the two principal directions 0° and 90°. Additionally, the temperature during the cure cycle is plotted. The data shows that the temperature significantly governs the strains in both directions due to the thermal expansion behavior of the metal and the laminate during cure.

Upon completion of the curing process and once the temperature returns to its starting value, the residual strains in the metal layer can be extracted from the data. The figure shows that in the end, positive residual strains form in the 0°-direction, while in the 90°-direction, negative strains prevail. Moreover, the residual strains in 0°-direction show higher absolute values than the ones in 90°-direction.

Figure 4.5 shows a similar picture for the modified cure cycle MOD1. The kink at the end of the MOD1 temperature curve is caused by opening the lab oven door to allow the laminate to cool to room temperature. Compared to Fig. 4.4, the final

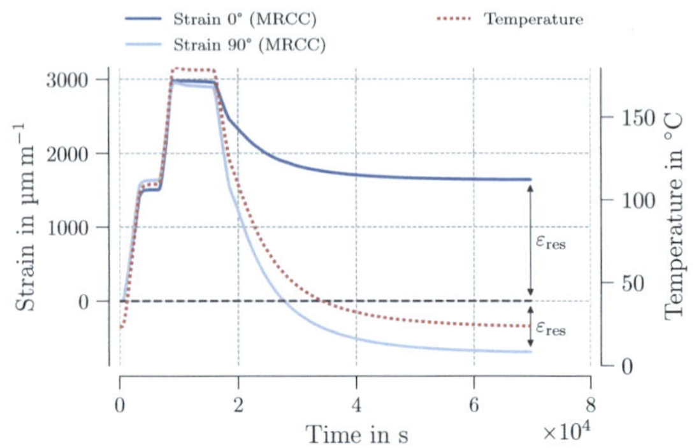

Fig. 4.4 Temperature and measured strains in the two principal in-plane directions (0° and 90°) of the metal layers during cure with the manufacturer-recommended cure cycle MRCC.

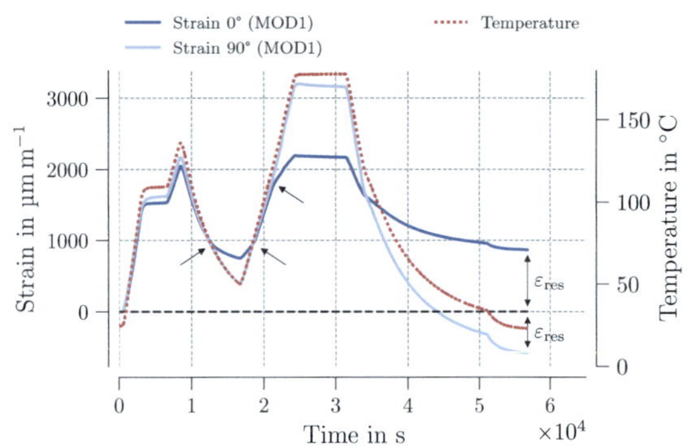

Fig. 4.5 Temperature and measured strains in the two principal in-plane directions (0° and 90°) of the metal layers during cure with the modified cure cycle MOD1.

residual strains in the metal are significantly smaller despite identical temperatures at this stage in the cure cycles.

When comparing the strains in the transverse direction (90°) with the strains in the fiber direction for both cure cycles, cure shrinkage strains at the maximum process temperature of 180 °C can be recognized. The strain data slightly decreases while the temperature remains constant. This behavior is more pronounced in the transverse direction, where the cure shrinkage strain of the resin is not constrained by the high stiffness of the fibers.

A similar behavior is observed for all modified cure cycles for the 0°-direction. However, all cure cycles result in different residual strains at the end of the process. Furthermore, several distinct knees (indicated by arrows in Fig. 4.5) can be identified within the recorded strains in the 0°-direction during the MOD cycles. Since the temperature shows a relatively constant gradient when the knees occur, they cannot be attributed to temperature changes.

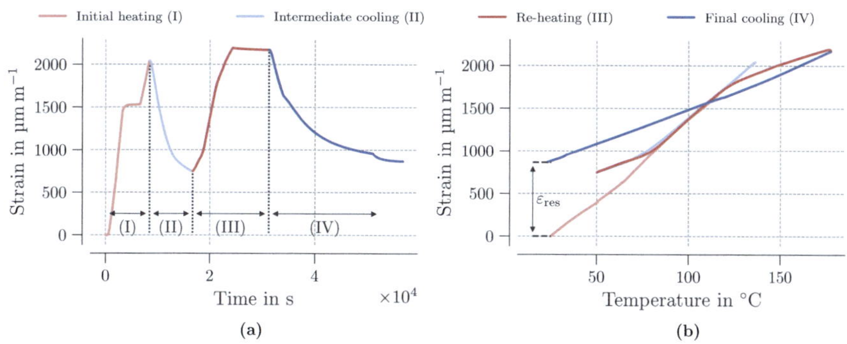

Fig. 4.6: **(a)** Strains over time and **(b)** strains over temperature for the MOD1 cure cycle. The characteristic stages (I)-(IV) of the cure cycle are differentiated by color. ε_{res} illustrates the final residual strains (adapted from [32]).

Curing Strains vs Process Temperature

For a detailed analysis of the curing strains over the process temperature, the recorded strains of MOD1 in 0° are used in the following. Plotting the strain data over the actual temperature is advantageous for a detailed interpretation. The strain data is divided into heating and cooling stages to find the way along the curve. These stages are highlighted in Fig. 4.6a. Figure 4.6b shows the strain data plotted against the temperature. In contrast to Fig. 4.6a, the residual strains at room temperature Fig. 4.6b are found on the left of the graph at the same temperature at which the experiment started. The kinks already observed in Fig. 4.5 can be attributed to specific temperatures in the strain temperature data where the slope of the curve significantly changes. It can also be seen that the behavior is relatively linear for most of the rest of the curve.

To get a detailed view of the strain data, the four curing stages are separately plotted in Fig. 4.7. Moreover, the linear segments of the curve are linearly interpolated, and the gradient is calculated. The gradient of the strain temperature data is interpreted as the instantaneous CTE of the substrate to which the strain gauge is bonded.

During the initial heating from starting temperature to 110 °C in the MOD1 cycle, the individual plies of the laminate behave independently of each other. Hence, the CTE is determined to be around $19 \times 10^{-6}\,K^{-1}$, which is precisely the CTE of the steel material used within the FML and given in Table 3.2.

During the intermediate cooling, the CTE shown in Fig. 4.7 is still in the range of the independent metal. Only before dropping below a temperature of about 75 °C the CTE decreases significantly from $18.03 \times 10^{-6}\,K^{-1}$ to a value of $8.15 \times 10^{-6}\,K^{-1}$, indicating a phase transition in the resin. This is attributed to the actual process temperature falling below the instantaneous glass transition temperature of the resin, which is accompanied by the transition into its glassy elastic state and a significant increase in the elastic resin modulus.

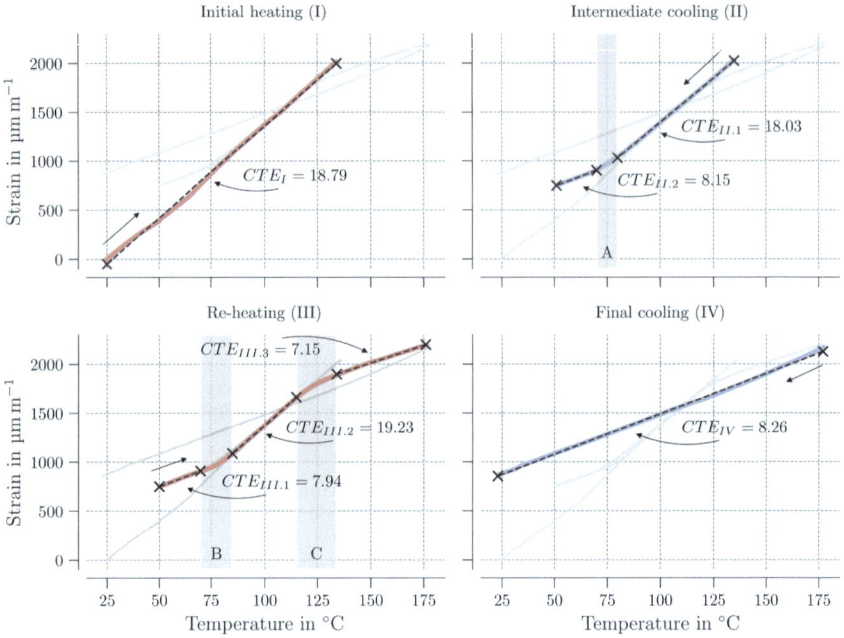

Fig. 4.7: The evaluation of the coefficients of thermal expansion for each stage of the MOD1 cure cycle indicates whether the free thermal expansion of the metal sheets or laminate behavior governs the strain recordings; the CTEs are given in $10^{-6}\,K^{-1}$. A: temperature falls below the instantaneous glass transition temperature, B: temperature exceeds the instantaneous glass transition temperature, C: the conversion exceeds the gel point (adapted from [32]).

After the subsequent re-heating to the final cure temperature of 180 °C, the CTE of the metal layer changes again from $7.94 \times 10^{-6}\,K^{-1}$ to $19.23 \times 10^{-6}\,K^{-1}$ at around the same process temperature. It is supposed that the resin transfers into its viscous state again, accompanied by a loss of the interaction between metal and fiber plies, which allows free thermal expansion of the metal. From this behavior, it can be assumed that the degree of cure has not yet reached a stage with a sufficient shear modulus to counterbalance the low thermal expansion of the fibers with the high expansion of the metal sheets. Thus, any interlaminar forces that led to a change in the slope during the previous cooling step are released after passing T_g again. The final interaction between the FML layers takes place at around 121 °C, which again changes the slope to approximately the final CTE of the laminate ($7.15 \times 10^{-6}\,K^{-1}$). The slightly smaller CTE compared to final cooling indicates that thermal expansion is partly compensated by cure shrinkage strains. From there, it can be assumed that the degree of cure has advanced to a stage with sufficient stiffness, where the laminate cannot change into a viscous state again despite a further increasing process temperature.

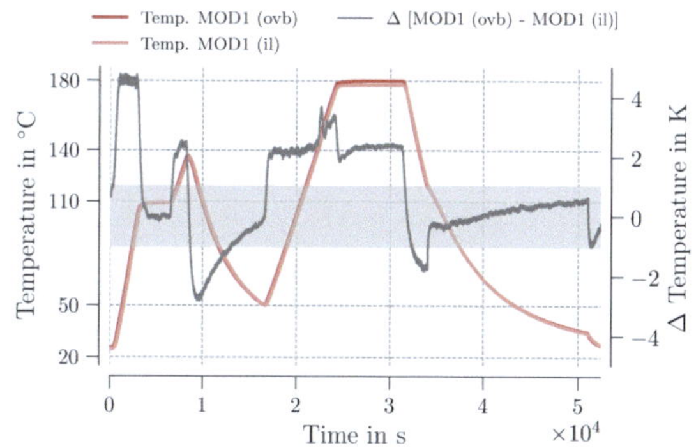

Fig. 4.8 Difference in temperature between the thermocouples placed inside laminate (il) and outside of the vacuum bag (ovb) for MOD1. For steep temperature gradients and during cure, a temperature difference of more than ±1 K (shaded area) is measured (adapted from [32]).

After the indissoluble bond between the individual layers of the FML has developed, the strain-temperature behavior is very much linear with the ultimate laminate CTE ($8.26 \times 10^{-6}\,K^{-1}$) during final cooling.

Large temperature gradients in the process cause non-linearities in the measurement data. This effect is seen for most cure cycles investigated because of several reasons. The main reason is the deviation in the temperatures of the measuring strain gauge and the dummy strain gauge. During the experimental setup, the dummy strain gauge is placed outside the vacuum bag in the cure oven. Figure 4.8 shows that for steep temperature ramps, the temperature between the compensation strain gauge and measurement strain gauge differs by up to 4.6 K for MOD1. Consequently, this leads to overcompensation of the temperature-induced strain during mathematical correction of the strain gauge signal. However, since the final residual strains are recorded at the same temperature as the starting temperature of the measurement, this influence will not affect the residual stress evaluation. Furthermore, evaluating distinct points in the strain signal will not be affected since the compensation error will only shift the strain reading vertically. Consolidation effects in the laminate stack during initial heating, temperature-dependent stiffness of the strain gauge adhesive and metal foil, and non-linearities in the CTEs of FRP and metal can potentially introduce additional non-linear effects. The compensation effect most significantly contributes to the non-linear strain signal for high-rate temperature changes.

Figure 4.9 shows the respective strain-temperature data of the modified cure cycles MOD2 and MOD3. The data of MOD2 is very similar to the strain data observed in MOD1. Only during intermediate cooling, with a cooling rate four times faster than in MOD1, does the temperature compensation introduce some errors. A completely different strain evolution is found for MOD3, despite the cooling temperature only being 5 K above that of MOD2. Here, the interaction between the single plies of the laminate already takes place before the intermediate cooling. Consequently, the strain behavior after reaching that point is linear for the remaining cure cycle. This can no longer be reversed once a certain degree of curing has been exceeded. The MOD3 cycle shows that the cooling temperature must be selected and controlled precisely to realize the bonding of the individual layers at the lowest possible temperature,

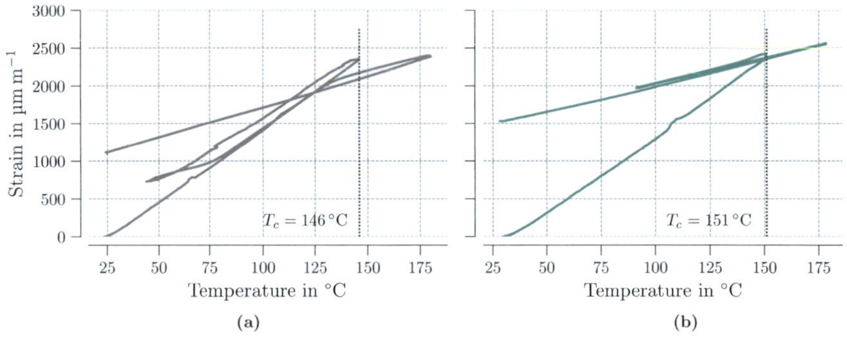

Fig. 4.9: Measured strains over the temperature of two cure cycles: **(a)** MOD2 and **(b)** MOD3. In the MOD3 cycle, bonding of the single layers takes place before intermediate cooling, which leads to significant differences in the strain signal despite only a small difference in the cooling temperature (T_c = 151 °C) compared to MOD2 (T_c = 146 °C).

an effect also shown in FBG measurements by [23]. This further shows that a homogeneous temperature distribution across the specimen thickness and over the size of the specimen must be ensured; otherwise, an inhomogeneous residual stress state may result after production.

4.2.2 Determination of Bonding Temperatures

Despite the final residual strains, the temperature at which bonding occurs within each cure cycle can also be determined from the cure strain data. This temperature is used as a reference temperature during residual stress calculations for that specific cure cycle in the literature [23]. While the measured residual strains are only valid for the specific layup, the bonding temperatures are independent of the layup but only depend on the cure cycle used.

The bonding temperature is defined as the point in the strain data, after which the slope behaves linearly until the end of the cure cycle. It is approximated by the highest change in the slope of the strain data before reaching that linear state. At this temperature, the transition from the thermally free expansion to the impeded expansion happens.

The strain data from the heating stage to the final cure temperature is derived twice with respect to temperature to determine this temperature for all the cure cycles. Because of the noise in the strain data, the first derivative is smoothed before the second derivative is calculated. The temperature of the global minimum of the second derivative is defined as the bonding temperature. This is illustrated exemplarily in Fig. 4.10 for the strain data of the MOD1 cure cycle. Moreover, the point at which the current process temperature exceeds the glass transition temperature ($T \geq T_g$) of

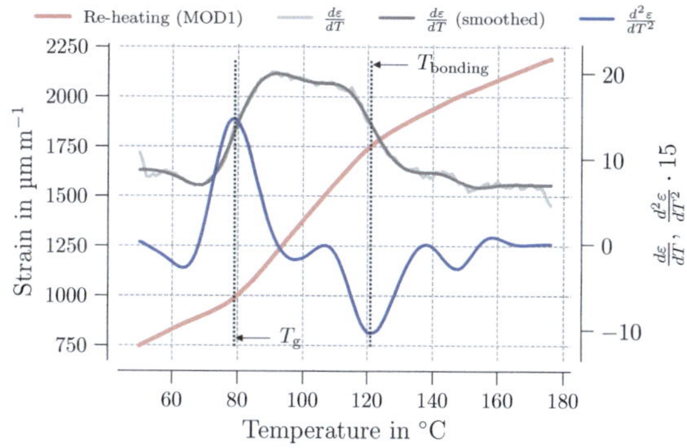

Fig. 4.10 Evaluation of the bonding temperature by derivation of the strain data during re-heating (exemplarily for MOD1). The minimum of the second derivative shows the transition from free linear expansion to constraint linear expansion of the metal sheet (adapted from [32]).

Table 4.1: Residual strains and bonding temperatures for each cure cycle and the two strain gauges. Residual strain and relative standard deviation (RSD) correspond to room temperature (20 °C).

	Final residual strain in µm m^{-1}			$T_{bonding}$ in °C	
	SG1	SG2	RSD in %	SG1	SG2
MRCC	1544	1590 [a]	2.1	172	172 [a]
MOD1	829	920	7.0	121	121
MOD2	1088	1064	1.6	129	129
MOD3	1441	1392	2.4	151	151

[a] values of second strain gauge (SG2) for MRCC derived from additional cure cycle

the resin during re-heating is indicated. This is accompanied by the transition from constraint to free thermal expansion.

Table 4.1 provides the final residual strains interpolated for room temperature (20 °C) and the bonding temperatures derived from the different cure cycles and strain gauges. Interestingly, although the final residual strains slightly differ within the two strain gauges applied to the same specimen, the bonding temperature is identical for all the strain gauges used in the same experiment. The bonding temperature for the MRCC is identical for even two different manufacturing runs. This indicates that the bonding temperature is a robust process-specific parameter.

4.2.3 Residual Stress Determination

To obtain residual stresses from measured strains, Eq. (4.3) converts the residual strains obtained at the end of each cure cycle into corresponding residual stresses. Since not all cure cycles end at the same temperature, a reference temperature of T_R = 20 °C is set for the comparison. The final cool-down step is linearly interpolated for all the strain gauge data, and a final laminate CTE is derived for each measurement.

Table 4.2: Measured residual strains and consequent residual stresses in the metal layers for all cure cycles and strain gauges for the two main in-plane directions. Residual strains are interpolated ($\varepsilon_{\mathrm{res}} = CTE\, T_{\mathrm{R}} + b$) for a reference temperature $T_{\mathrm{R}} = 20\,°\mathrm{C}$. Residual stresses are calculated using Eq. (4.3).

Cure cycle / strain gauge / orientation	CTE in $10^{-6}\,\mathrm{K}^{-1}$	b in $\mu\mathrm{m\,m}^{-1}$	$\varepsilon_{\mathrm{res}}$ in $\mu\mathrm{m\,m}^{-1}$	$\sigma_{\mathrm{res,\ meas.}}$ in MPa
MRCC-SG1 (0°)	8.09	(-)[a]	-	
MRCC-SG2 (0°)	8.57	1372.90	1544	275
MRCC2-SG1 (0°)	8.36	1422.96	1590	284
MOD1-SG1 (0°)	8.26	664.25	829	131
MOD1-SG2 (0°)	8.51	750.26	920	148
MOD2-SG1 (0°)	8.78	912.64	1088	183
MOD2-SG2 (0°)	8.02	903.76	1064	178
MOD3-SG1 (0°)	(6.89)	1303.51	1441	254
MOD3-SG2 (0°)	(7.14)	1249.18	1392	244
Mean	8.37			
MRCC (90°)	23.23	-1250.40	-786	-68
MRCC2 (90°)	22.63	-1254.18	-802	-68
MOD1 (90°)	23.98	-1162.08	-682	-91 / -85
MOD2 (90°)	-	-	(-715)[b]	-82 / -83
MOD3 (90°)	-	-	(-769)[b]	-71 / -74
Mean	23.28			

[a] not meaningful due to jumps of the signal, [b] interpolated

Table 4.2 gives each measurement's corresponding final CTE values, the corresponding strains at $T_R = 20\,°\mathrm{C}$ and the consequent residual stresses. The results show that the laminate CTE is independent of the cure cycle and is determined in the range of 8 to $8.8 \times 10^{-6}\,\mathrm{K}^{-1}$ in the longitudinal direction and 23.2 to $24 \times 10^{-6}\,\mathrm{K}^{-1}$ for the transverse direction. Only the final CTEs of the specimen manufactured with cure cycle MOD3 deviate from all other measurements. The specimen is, therefore, excluded from the calculation of the mean CTE.

Comparing the residual stresses in the longitudinal direction of the metal ply with the tensile strength of the metal material ($\sigma_{ult} = 1121\,\mathrm{MPa}$, cf. Table 3.1) shows that the residual stresses already account for about 20% of the material strength when using the MRCC. By using modified cure cycles, the residual stress is reduced by up to 50% (compare MRCC to MOD1). The residual stresses in the 90°-direction are significantly smaller due to the uni-directional fiber architecture within the investigated layup.

To discuss the question of whether the bonding and stress-free temperatures can be used interchangeably, Fig. 4.11 provides the measured residual stresses from Table 4.2 as a function of the bonding temperatures. It can be seen that with an increasing bonding temperature, the final residual stresses also increase. Figure 4.11 additionally provides the analytically calculated residual stresses using the CLT and the mean laminate CTEs found in Table 4.2. For the calculated residual stresses, the

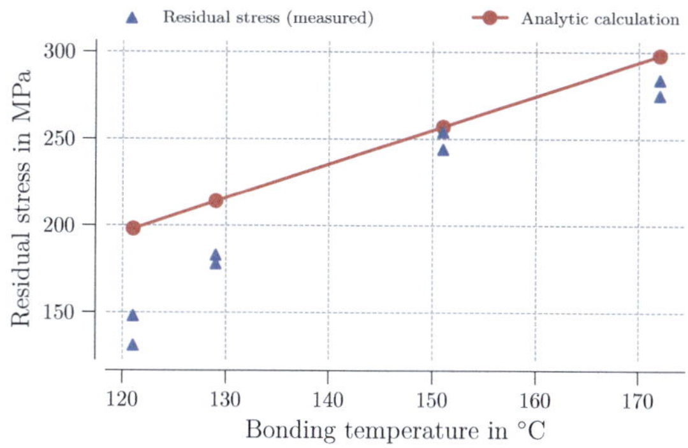

Fig. 4.11 Comparison of residual stresses in 0°-direction of the metal sheets as a function of the bonding temperature. For high bonding temperatures, the residual stress from measured strains agrees with the analytic solution under the assumption in Eq. (4.4). For lower bonding temperatures, this assumption results in larger differences.

following assumption was made:

$$T_{\text{sf, calc.}} = T_{\text{bonding}} \tag{4.4}$$

Figure 4.11 shows that for high bonding temperatures (151 K and 172 K) the measured and calculated values agree very well. However, higher deviations between measurement and calculation exist for the lower bonding temperatures of MOD1 and MOD2. Therefore, it can be concluded that assuming the stress-free temperature to be equal to the bonding temperature, Eq. (4.4), can lead to errors if the bonding temperature, determined with integrated sensors, and the final curing temperature differ significantly.

One suspected reason is the time difference between the gel point and the vitrification. For the MRCC and high bonding temperatures, there is only a short period of time between the gel point and vitrification, as the curing reaction takes place very quickly. With MOD cycles, on the other hand, the material is explicitly cooled after reaching the gel point and then slowly reheated to the curing temperature. As a result, there is a much more extended period of time between the gel point and vitrification. The material is in its visco-elastic state during this time, and stresses can be partially relaxed over time. In contrast, when considering the case of pure linear thermo-elasticity and perfect instantaneous bonding of the single layers in the laminate, stress-free and the bonding temperature would be identical.

4.3 Validation of Strain Gauge Measurements With Fiber Bragg Grating Sensors

To validate the strain gauge measurements, a specimen manufactured within the MOD2 cure cycle is additionally equipped with a fiber Bragg grating (FBG) sensor. An FBG sensor is an optical glass fiber with a Bragg grating engraved at a certain position, the measuring position. A laser light source sends a broad-spectrum light

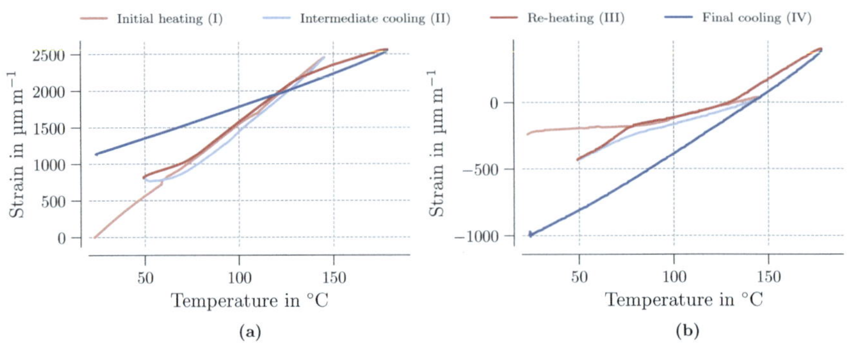

Fig. 4.12: Strains measured during cure with the MOD2 cure cycle with two different sensors: **(a)** Strain gauge attached to the top metal sheet and **(b)** FBG sensor embedded into the CFRP layer below the top metal sheet. The comparison of the strain signals shows the non-linear behavior of the FBG sensor during initial heating.

impulse into the glass fiber. The Bragg grating reflects a particular wavelength depending on the spacing of the grating, which is detected by a spectrometer. When the spacing of the grating is changed due to a mechanical or thermal strain, the wavelength of the reflected light λ_b changes accordingly. This shift in reflected wavelength $\Delta\lambda_b/\lambda_b$ can be directly converted into a corresponding strain ε of the fiber:

$$\varepsilon = \frac{\frac{\Delta\lambda_b}{\lambda_b} - C_T\Delta T}{C_S} \tag{4.5}$$

The thermo-optic coefficient C_T accounts for any wavelength shifts due to temperature-induced changes in the refractive index, and the photo-elastic coefficient C_S is a material constant of 0.78. This equation is valid for an FBG where no shear lag between the sensor and substrate occurs, which can be ensured by a sufficient bedding length of the glass fiber and a short length of the Bragg grating. [7, 16]

In this work, a 1550 nm fiber Bragg grating sensor with a grating length of 2.5 mm and a MIOPAS interrogator (SPK 155), which acts as the light source and spectrometer, are used. The bedding length of the FBG is 125 mm. The FBG sensor is embedded into the middle of the CFRP stack below the top metal layer. The FBG signal is recorded in parallel to the strain gauge signal. For the temperature compensation of the FBG signal, a thermo-optic coefficient C_T of $4.2 \times 10^{-6}\,\mathrm{K}^{-1}$ was assumed, which is in accordance with the literature [16].

Figure 4.12 shows the strains measured with the strain gauge and the FBG sensor within the same laminate. Significant differences are observed. During the initial heating stage (I) of the cure cycle, the strain measured with the FBG is highly non-linear. This is because the FBG can only measure the expansion of the fiber layers when the polymer possesses substantial shear modulus to force the FBG glass fiber to strain together with the substrate material. In contrast, the strain gauge measures the CTE of the metal sheets which results in linear strains in this stage of the cure

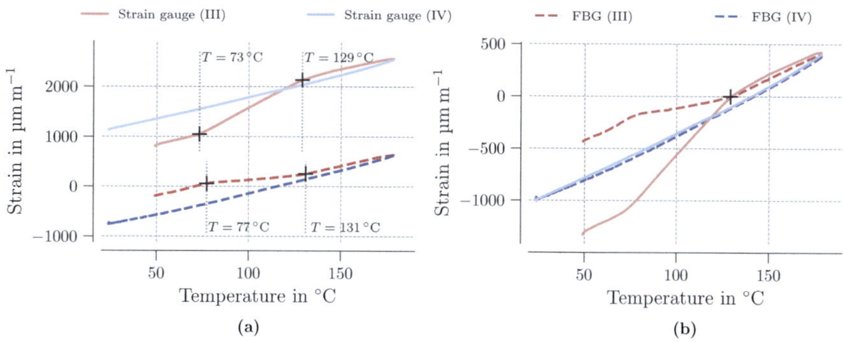

Fig. 4.13: Comparison of measured strain gauge and FBG strains during re-heating (III) and final cooling (IV) of MOD2 (strain gauge applied to the metal layer, FBG embedded into the fiber layer): **(a)** recorded strains and **(b)** strains zeroed at temperature $T = 129$ °C (adapted from [32]).

cycle. At the end of the cure cycle, the FBG sensor measures negative residual strains in contrast to the strain gauge. This indicates the compressive residual state of the fiber layers after manufacturing. However, due to the non-linear first part of the strain signal, these residual strains do not directly correlate with the compressive residual stress state in the fiber layers.

To compare the strains between strain gauge and FBG quantitatively, Fig. 4.13 zooms into the cure cycle stages (III) and (IV). Again, the re-heating (III) is shown in red, whereas the final cooling (IV) is plotted in blue. The strains indicate the transition from below the glass transition temperature to a renewed independent expansion state, which culminates in the final bonding of the individual layers. In contrast to the strain gauge data, the FBG signal transitions into a thermal expansion behavior with a smaller gradient at 77 °C. At this temperature, the resin transitions into its viscous state, accompanied by a loss of the connection between the different FML layers. Hence, the thermal expansion behavior of the sensor and the fiber ply is dominant, with both being smaller than the metal. Moreover, the FBG data shows a non-linear sensor behavior during the following stage, which is attributed to the interference of temperature-induced elongation of the sensor and the fiber ply and cure shrinkage strains caused by the resin. Only at 131 °C is the gradient increasing again to almost the same value as during the final cool-down. This indicates laminate behavior from this point onward, with the single FML plies bonded together.

A small deviation exists when evaluating the prominent temperatures by calculating the second derivatives. However, the temperature of the final bonding deviates only by 2 K from each other. This deviation is considered relatively small, given that two different measuring systems with different sensors in two different layers of the FML were used for the data collection.

In Fig. 4.13b, the strain data is zeroed at the strain gauge bonding temperature (129 °C) for both sensor types. Almost identical behavior in the strain data of both sensors can be seen from this point onward, which indicates linear elastic behavior of the entire laminate.

As stated in the beginning, the FBG measurements cannot directly quantify the residual stress state. However, the FBG data observe the same laminate behavior during the re-heating stage and the final cooling of the modified cure cycle. Furthermore, the distinct points where the strain signal gradient changes significantly can also be found in the FBG data.

4.4 Estimation of the Validity of Single Point Measurements

For both the measurements with strain gauges and the measurement with FBG, the strains were only measured at a few selected points in the laminate. The measured strains or bonding temperatures were then used to calculate the residual stress state of the entire laminate. This raises the question of whether the measurements taken at individual points are representative of the entire laminate. To discuss this question, a differentiation between a stress inhomogeneity in-plane (within a laminate layer) and an out-of-plane (laminate thickness direction) inhomogeneity is made.

Regarding residual stresses in the laminate plane, the literature shows that stress gradients can occur around the free edges of the laminate [21]. However, these stress gradients are only significant in a small area away from the edge, corresponding to approximately twice the thickness of the laminate. In-plane residual stress gradients can also be caused by temperature gradients during cure. These temperature gradients can result from complex airflow in the autoclave. For the small plate geometries investigated in this work and the manufacturing equipment used, the maximum temperature difference within the area where the tool is positioned in the autoclave is below 1 K throughout the entire cure cycle. Therefore, no significant difference in the residual stress state of the in-plane residual stresses is expected for the size and geometry of the laminates produced and tested. However, this could be of greater importance for industrial production processes with fully loaded autoclaves, extensive part geometries, steep heating gradients, nitrogen-filled pressure chambers, and manufacturing tools with high thermal mass [17, 24, 30]. Temperature differences of up to 76 K are reported in the literature for different positions on a large manufacturing tool during autoclave cure [12].

On the contrary, the out-of-plane residual stress state is mainly driven by the temperature distribution across the laminate thickness. A temperature gradient leads to differences in the cure rate and finally to a residual stress gradient through the thickness [1, 8, 34]. This phenomenon gets more relevant with increasing part thickness, and the heating and cooling ramps in the manufacturing process need to be adapted [20, 25]. For the out-of-plane temperature distribution, temperature differences of more than 20 K are reported [1, 35].

In this work, the temperature influence on the out-of-plane residual stress state is particularly interesting for the MOD cycles. The strain gauge measurements have shown that the resin cure kinetics is very sensitive to the temperature around the

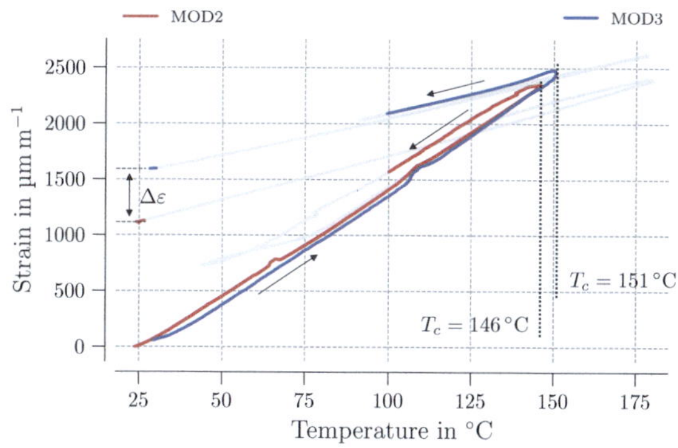

Fig. 4.14 The final residual strains in a laminate are very sensitive to small deviations in the process temperature. A difference of 5 K in the cooling temperature of the MOD cycle can lead to a significant difference in final residual strains.

intermediate cooling step in the MOD cycles. Figure 4.14 illustrates this finding. A temperature difference of 5 K of the cooling temperature T_c can significantly affect the final residual strains. This is in good agreement with [23] that showed that a temperature difference of 5 K when cooling is initiated or a difference in the heating or cooling ramps of $2\,\mathrm{K\,min^{-1}}$ can make a significant difference in the resin cure kinetics.

Therefore, this section will further investigate the temperature distribution across the laminate thickness to ensure that the previous cure strain measurements and the consequent residual stress calculation are representative of the entire laminate. The focus will be on the MOD cycles since the influence is assumed to be more significant here due to the steep temperature gradients during intermediate cooling and their particular sensitivity to temperature deviations.

4.4.1 Experimental Temperature Distribution Across the Thickness in FML

For the temperature measurements, thermocouples (TCs) of type K are placed in the center of the laminates during the layup. Table 4.3 gives the positions with respect to the laminate thickness for specimen ID3a. Additionally, the positions of the TCs in a monolithic CFRP laminate are given, which is of interest in the next section. The positions of the TCs in the laminates are defined based on preliminary assumptions on the temperature distribution. The temperature distribution across the laminate thickness is measured during the manufacturer-recommended and modified cure cycles for the layups ID2, ID3a, and ID6b to cover a wide range of laminate thicknesses. The highest deviations in temperature are found for both cure cycles in the layup ID3a. This laminate possesses the largest thickness (6.52 mm) of all FML layups manufactured.

The measured temperatures over time throughout the MOD cycle for the FML specimen are shown in Fig. 4.15. Additionally, the temperatures of the tool and

Table 4.3 Position of TCs, in respect to the laminate thickness, from 0% (bottom = tool side) to 100% (top = vacuum bag side).

	TC1	TC2	TC3	TC4
ID3a	6%	35%	65%	94%
CFRP	3%	26%	50%	82%

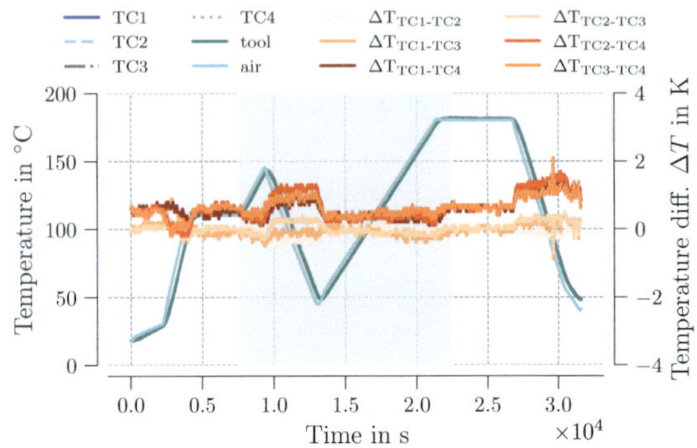

Fig. 4.15 Temperatures measured during the modified cure cycle and layup ID3a within the autoclave and corresponding temperature difference between all the thermocouples placed inside the laminate at different positions according to Table 4.3 (adapted from [31]).

autoclave air are shown. The tool and TC temperatures show only minor differences, so the curves can hardly be differentiated in the figure. In contrast, the air temperatures always precede the temperatures measured inside the laminate, as the autoclave regulates the air temperature, which then circulates around the tool and the vacuum bag enclosing the laminate. The maximum deviation between all the thermocouples inside the FML is below 2 K during the final cool-down step in the cure cycle. However, as the strain gauge measurements show, the residual stress state is already determined in this stage of the cure cycle. When looking at the relevant stage of the cure cycle (shaded area), the highest deviation is found to be 1.3 K during the intermediate cooling step. The deviations for the other layups (ID2 and ID6b) are smaller. Compared to the 5 K difference discussed in [23] and measured in this work (Fig. 4.14), the difference is assumed to be too small to result in significant deviations of the residual stress state across the laminate thickness. Therefore, the development of a homogeneous residual stress state in all investigated FMLs (up to a thickness of 6.52 mm) can be assumed.

4.4.2 Numerical Modeling of the Through-Thickness Temperature Distribution in FRP Laminates

This section further investigates the temperature distribution across the laminate thickness in a numerical model[3] to understand better and contextualize the large temperature deviations reported in the literature.

[3] The numerical heat transfer simulation was developed in cooperation with Tetiana Pittsyk from the Institute of Mechanics and Adaptronics of the Technische Universität Braunschweig.

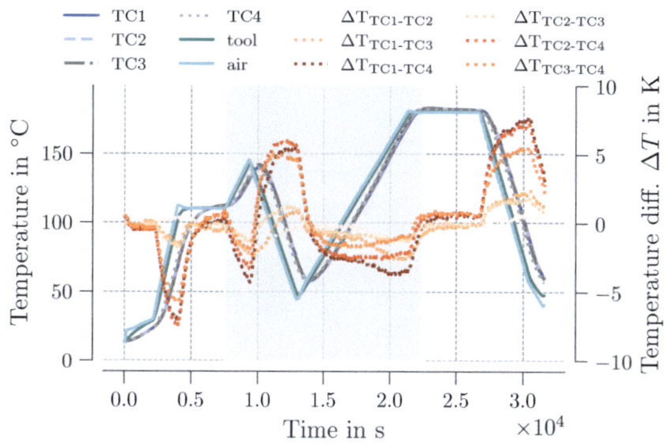

Fig. 4.16 Temperature distribution inside a 20 mm thick monolithic CFRP laminate during autoclave cure with the MOD cycle and corresponding temperature differences between the TC positions (adapted from [31]).

Since the heat during cure in an autoclave is transferred only from the top and bottom of the laminate, a temperature gradient depends on the heat transfer characteristics in the laminate thickness direction. The heat transfer is governed by the laminate's specific heat capacity, thermal conductivity, and thickness. CFRP possesses a lower thermal conductivity, especially in the direction of the laminate thickness, compared to steel (cf. Sec. 3.2). Therefore, an additional monolithic CFRP laminate equipped with thermocouples is manufactured to increase the temperature difference and validate the numerical model. The monolithic CFRP laminate comprises 154 prepreg plies and possesses a thickness of 20 mm. The laminate is manufactured using the MOD cure cycle since higher temperature deviations are expected due to the fast intermediate cooling step.

The temperatures measured during manufacturing are plotted in Fig. 4.16. Significantly higher temperature deviations between the different thermocouples are measured than for the thinner FML specimens. It can be seen that the temperature in the laminate differs by around 5 K during intermediate cooling between the thermocouple being furthest away from the tool (TC4) and the other three thermocouples (TC1, TC2, TC3).

The experimental results of the CFRP laminate are used to validate a numerical simulation. The simulations of the temperature distribution are performed using an Abaqus heat transfer analysis. A 3D model of the laminate, including the steel tool, is created and meshed with 20-node quadratic brick elements (DC3D20) to extend the model for more complex part geometries in the future, and since computational efficiency is not a concern. All the elements in the model are given a temperature of 293 K in the initial step in the form of a predefined field.

For the transient step, several simplifications are made. The thermal material parameters (thermal conductivity and specific heat capacity) are only assumed to be temperature-dependent. Although the dependence on the degree of cure is reported in several works, e.g., [3], it is neglected to reduce the modeling effort. From the formulas derived by [8] and provided in Sec. 3.2, it can be found that the influence of the degree of cure is significantly smaller than the temperature influence. Therefore, this simplification in the model seems acceptable. Furthermore, the temperature in

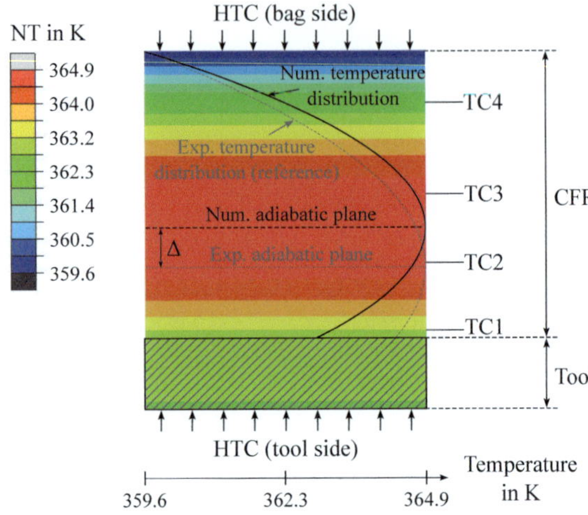

Fig. 4.17 Cross-sectional view of the numerical model with nodal temperature (NT) distribution during the intermediate cooling step at the time increment 11 624 s. The experimental temperature distribution added as a reference shows the error made with the initially assumed HTCs (adapted from [31]).

the transient step is applied as a surface film condition with the autoclave temperature defined through an amplitude for the cure cycle.

Slesinger et al. [26] show that the heat transfer coefficient (HTC) for the circulating hot air depends on the location inside the autoclave, as well as on the actual process temperature and pressure. For the investigated setup, the HTCs varied between $34\,\mathrm{W\,m^{-2}\,K^{-1}}$ and $186\,\mathrm{W\,m^{-2}\,K^{-1}}$. A value of $100\,\mathrm{W\,m^{-2}\,K^{-1}}$ is observed for most of the cure cycle around the area where the tool with the laminate is placed. Although the autoclave used in this work differs, the value is used in the numerical model on the top ply of the laminate and the bottom of the steel tool as an initial starting value. To simplify the simulation, the variation of the HTCs throughout the cure cycle is neglected. The other surfaces of the laminate are assumed to be adiabatic.

The initial results show that the temperature distribution is qualitatively correct, but there is a significant difference between the experiment and numerical results. Figure 4.17 illustrates this on a cross-section of the numerical model and the nodal temperature (NT) distribution over the thickness at a selected time increment (t=11 624 s) during the intermediate cooling step in the MOD cycle. The figure further shows the tool, the thermocouple positions, and the adiabatic plane in the laminate. The adiabatic plane defines the position in the thickness direction across which no heat is transferred. The adiabatic plane is not in the symmetry plane of the laminate, although the layup is symmetric because, on the bottom side, the heat is transferred through the tool. On the top side, it is directly transferred to the laminate. Compared to the experimental adiabatic plane, the numerical adiabatic plane is shifted towards the center of the laminate. This leads to the conclusion that more heat is transferred through the tool side in the numerical model compared to the experiment. Hence the initially assumed values for the HTCs ($100\,\mathrm{W\,m^{-2}\,K^{-1}}$) from the literature do not represent the experimental setup used in this work. Therefore, the HTCs are iteratively updated until the numerical temperature distribution and adiabatic plane match the experimental results.

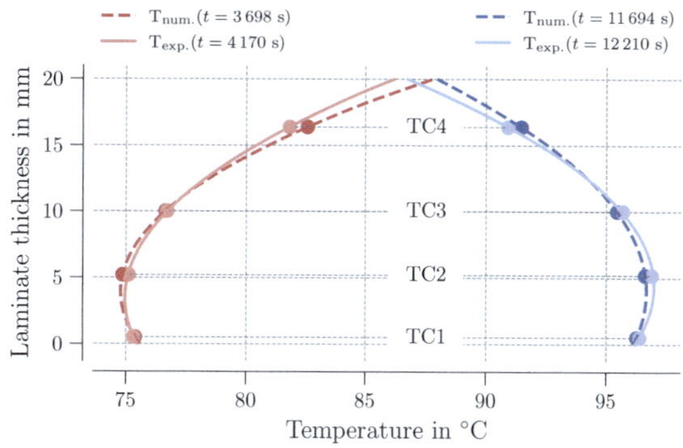

Fig. 4.18 Experimental and numerical (updated HTCs) temperature distribution across the thickness of the 20 mm CFRP laminate, with data points indicating the TC locations and lines representing the parabolic fit (adapted from [31]).

The updated model results in HTCs of $80\,W\,m^{-2}\,K^{-1}$ at the top of the laminate and $40\,W\,m^{-2}\,K^{-1}$ at the bottom of the tool. With these values, the experimental temperature distribution across the thickness of the laminate is very well represented in the numerical model. This is shown in Fig. 4.18, where the temperatures of the four TC positions are plotted over the laminate thickness for selected time increments during the initial heating and the intermediate cooling stage in the cure cycle. Slight differences in the time increments between the experiment and the simulation originate from the different increment sizes but do not affect the comparability. The increments were chosen so that the experimental and numerical temperatures at the position TC1 are identical. Based on the results of [25], the temperature distribution across the thickness can be assumed to follow a second-order polynomial at steady-state. Fitting a quadratic function through the data points by minimizing the sum of the squared residuals still demonstrates an appropriate fit for the experimental and numerical results, although the heat transfer is not steady at this stage. The vertex of the corresponding parabola defines the adiabatic plane in the laminate. A very good agreement between experimental and numerical results is achieved during heating and cooling.

In Fig. 4.19, the experimental and numerical temperature differences between all TC positions are plotted over the cure cycle. Additionally, the absolute temperature of the numerical TC1 is given as a reference. The results show that the numerical model can accurately represent the temperature distribution measured during the experiment over the entire cure cycle, despite the simplifications made and the non-consideration of the cure kinetics in the model.

To evaluate the temperature distribution of more than 20 K reported in the literature [1, 35], the validated model is used to extrapolate the results for even thicker laminates. The model is used to additionally simulate a 40 mm thick CFRP laminate, a laminate thickness that is common at the wing root joint of an aircraft [13]. The temperatures throughout the cure cycle are shown in Fig. 4.20 for the top surface and adiabatic plane of the laminate. The results indicate an expected temperature difference of up to 20 K between the two locations in the critical process window (shaded area). Furthermore, the temperature at the point in the MOD cycle when

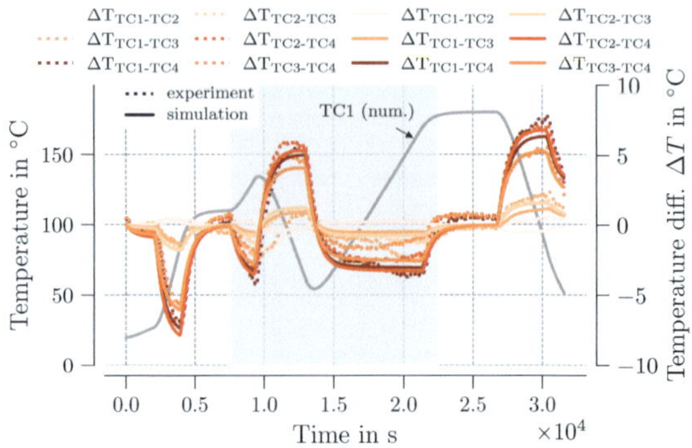

Fig. 4.19 Comparison of the numerical and experimental temperatures at the TC positions inside the 20 mm thick CFRP laminate during the MOD cycle (adapted from [31]).

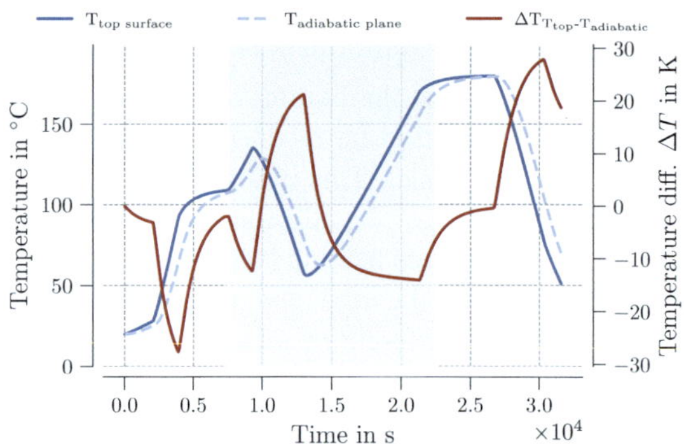

Fig. 4.20 Numerical temperature distribution during MOD cycle of a 40 mm thick CFRP laminate shows significant differences between the top laminate surface and the adiabatic plane in the laminate (adapted from [31]).

curing is initiated significantly deviates between the two positions in the laminate. These deviations can lead to considerable differences in the cure kinetics of the resin, leading to different bonding temperatures between the plies and, hence, to substantial differences in the residual stress state. However, for these very thick composite laminates, the exothermic reaction of the resin might not be neglected anymore and can further influence the results.

The numerical results show that temperature differences of up to 20 K are realistic for thick laminates. This, in consequence, limits the validity of single-point measurements for residual stress quantification in those laminates. Therefore, the homogeneity of the temperature across the laminate thickness is crucial for residual stress quantification based on a single measurement location.

4.5 Conclusion and Discussion

With the strain gauge method presented, the residual strains in the metal layers of an FML can be determined with respect to the beginning of the cure cycle. In the investigated layup, the residual stresses account for about 20% of the material strength when using the MRCC. By using MOD cycles in the laboratory environment, the residual stresses can be reduced by up to 50% (compare MRCC to MOD1).

The strain signal recorded on top of the metal sheets during the curing processes clearly shows the stages in the cure cycle in which interaction between the metal layers and the FRP layers takes place. Due to the significantly lower CTE of the fiber layers in the fiber direction compared to the metal sheets, the measured CTE transitions from the metal's free thermal expansion CTE to a reduced CTE, governed by the constraint of the fiber layers.

When this interaction between fiber layers and metal sheets is non-reversible due to a sufficient degree of cure of the resin material, the corresponding process-specific temperature at which the non-reversible connection occurs is called the bonding temperature. When the degree of cure of the resin is still small, but the process temperature falls below the instantaneous glass transition temperature of the polymer, the polymer transitions into its glassy state. This is associated with a significant increase in the elastic modulus of the polymer and can, therefore, also be identified in the strain signals. Since this is a reversible process, the temperature associated with this effect differs from the bonding temperature.

The results further show that assuming the stress-free temperature equals the bonding temperature leads to errors if the bonding temperature and the maximum process temperature differ significantly. One possible reason for this is the occurrence of non-linear effects, particularly pronounced when there is a long period between the bonding temperature and the final process temperature. If the individual layers of the FML interact early in the production process, stresses can be reduced by flow and relaxation processes due to the visco-elastic material behavior of the polymer matrix before vitrification. Therefore, determining the stress-free temperature for laminates manufactured in MOD cure cycles is of particular importance for residual stress calculation when absolute strain measurements are not available.

The investigations regarding the temperature distribution across the laminate thickness revealed that single-point measurements are sufficient to quantify the residual stress state within the entire laminate for the laminates investigated in this work. However, this correlation is no longer valid for thick laminates, and a residual stress gradient across the laminate thickness will occur. Analysis of the temperature distribution across the sample thickness also revealed that the large temperature gradients used during MOD cycles cannot be applied uniformly to all layers, making the required temperature control impossible for thick laminates.

With regard to the research hypothesis formulated for this chapter, *in contrast to FRPs, strain sensors attached to the metal layers of FMLs enable the recording of meaningful time-based strain data from the beginning of the manufacturing process*, the results show that this hypothesis can be confirmed. The presented strain gauge method allows for quantifying residual stresses from the beginning of manufacturing, which is impossible with FBG sensors embedded into the fiber material since they only attach to the substrate during manufacturing. Therefore, when evaluating the FBG strains, there is no starting value for calculating the resulting residual stresses. A significant temperature in the process is the bonding temperature, which needs to be used as a reference temperature instead, which is often done in the literature. However, the strain gauge measurements have shown that the bonding temperature from measured strain data needs to be differentiated from the stress-free temperature. The error is small if the stress-free temperature is close to the final curing temperature. However, this is not the case for modified cure cycles.

Furthermore, despite measuring the strains in the metal, which does not change its elastic state during cure in contrast to the polymer, the interaction between metal and FRP layers still leads to changes in the strain gradient during temperature ramps in the cure cycle. The polymer is only able to transfer significant amounts of loads after passing the gel point, which comes along with a measurable elastic modulus. Likewise, when the process temperature is decreased below the actual glass transition temperature of the polymer, the elastic modulus abruptly increases, which can also be seen in the strain signal. Hence, the strain signal indicates the resin state of cure. However, it is not easy to derive direct statements about the degree of cure or the current glass transition temperature.

The strain gauge method requires careful evaluation of its applicability beyond laboratory conditions. Challenges are associated with measuring residual stress using strain gauges in industrial manufacturing processes. First, the strain gauges must be attached to the corresponding layers with a high-temperature-resistant adhesive before the laminate manufacturing process. The adhesive must cure at temperatures that exceed the maximum temperature used in laminate manufacturing, which results in an additional process step. Unlike the manufacturing processes considered here, a cover plate is often added, making it more challenging to attach the strain gauges. Three cables and at least one temperature sensor for each strain gauge must be routed out of the vacuum bagging and out of the oven or autoclave, significantly complicating the manufacturing setup. The solder points at the connection between the cable and the strain gauge must be positioned carefully to ensure they do not contact the metallic surface or neighboring solder points, even if the solder softens under pressure and temperature. Ideally, temperature measurements should be taken close to each strain gauge. The sensitive glass body, along with the applied dummy strain gauge, must also be placed in the autoclave or oven, with the corresponding cables routed outside.

In summary, while the method is effective for understanding a material system and identifying improvements for reducing residual stress, it appears less suitable for determining residual stress during industrial manufacturing processes. Therefore, the next chapter discusses a method that uses asymmetric layups for residual stress quantification, which seems better suited for industrial manufacturing.

References

1. Bogetti, T.A., Gillespie, J.W.: Process-induced stress and deformation in thick-section thermoset composite laminates. Journal of Composite Materials **26**(5), 626–660 (1992). doi:10.1177/002199839202600502

2. Canal, L.P., Benavente, M., Hausmann, M., Michaud, V.: Process-induced strains in RTM processing of polyurethane/carbon composites. Composites Part A: Applied Science and Manufacturing **78**, 264–273 (2015). doi:10.1016/j.compositesa.2015.08.018
3. Chern, B.C., Moon, T.J., Howell, J.R.: Measurement of the temperature and cure dependence of the thermal conductivity of epoxy resin. Experimental Heat Transfer **6**(2), 157–174 (1993). doi:10.1080/08916159308946451
4. Crasto, A.S., Kim, R.Y., Russell, J.D.: In situ monitoring of residual strain development during composite cure. Polymer Composites **23**(3), 454–463 (2002). doi:10.1002/pc.10447
5. Daniel, I.M., Liber, T.: Effect of laminate construction on residual stresses in graphite/polyimide composites. Experimental Mechanics **17**(1), 21–25 (1977). doi:10.1007/BF02324267
6. Daniel, I.M., Liber, T., Chamis, C.C.: Measurement of Residual Strains in Boron-Epoxy and Glass-Epoxy Laminates. In: E.M. Wu (ed.) Composite Reliability, pp. 340–340–12. ASTM International, 100 Barr Harbor Drive, PO Box C700, West Conshohocken, PA 19428-2959 (1975). doi:10.1520/STP32318S
7. Hill, K.O., Meltz, G.: Fiber Bragg grating technology fundamentals and overview. Journal of Lightwave Technology **15**(8), 1263–1276 (1997). doi:10.1109/50.618320
8. Johnston, A.A.: An integrated model of the development of process-induced deformation in autoclave processing of composite structures. Dissertation, University of British Columbia, Vancouver, Kanada (1997). doi:10.14288/1.0088805
9. Kang, H.K., Kang, D.H., Bang, H.J., Hong, C.S., Kim, C.G.: Cure monitoring of composite laminates using fiber optic sensors. Smart Materials and Structures **11**(2), 279–287 (2002). doi:10.1088/0964-1726/11/2/311
10. Kappel, E., Prussak, R., Wiedemann, J.: On a simultaneous use of fiber-Bragg-gratings and strain-gages to determine the stress-free temperature Tsf during GLARE manufacturing. Composite Structures **227** (2019). doi:10.1016/j.compstruct.2019.111279
11. Kim, Y.K., Daniel, I.M.: Cure Cycle Effect on Composite Structures Manufactured by Resin Transfer Molding. Journal of Composite Materials **36**(14), 1725–1743 (2002). doi:10.1177/0021998302036014598
12. Kluge, N.E., Lundström, T.S., Ljung, A.L., Westerberg, L.G., Nyman, T.: An experimental study of temperature distribution in an autoclave. Journal of Reinforced Plastics and Composites **35**(7), 566–578 (2016). doi:10.1177/0731684415624768
13. Koord, J.: Design methodology for highly loaded composite bolted joints with local metal hybridization at low temperature: DLR-Forschungsbericht - DLR-FB-2024-11. Dissertation, German Aerospace Center, Braunschweig (2024). doi:10.57676/mzxc-3158
14. Lawrence, C.M., Nelson, D.V., Spingarn, J.R., Bennett, T.E.: Measurement of process-induced strains in composte materials using embedded fiber optic sensors (1996). doi:10.2172/226060
15. Litoš, P., Švantner, M., Honner, M.: Simulation of Strain Gauge Thermal Effects During Residual Stress Hole Drilling Measurements. The Journal of Strain Analysis for Engineering Design **40**(7), 611–619 (2005). doi:10.1243/030932405X30812
16. Luyckx, G., Voet, E., Lammens, N., Degrieck, J.: Strain measurements of composite laminates with embedded fibre bragg gratings: criticism and opportunities for research. Sensors (Basel, Switzerland) **11**(1), 384–408 (2011). doi:10.3390/s110100384
17. Maffezzoli, A., Grieco, A.: Optimization of Parts Placement in Autoclave Processing of Composites. Applied Composite Materials **20**(3), 233–248 (2013). doi:10.1007/s10443-012-9265-8
18. Michigan Scientific Corporation: Technical Note No. 105: Strain gage output in response to free thermal expansion, non-thermal stress, and thermal stress
19. Mulle, M., Collombet, F., Olivier, P., Grunevald, Y.H.: Assessment of cure residual strains through the thickness of carbon–epoxy laminates using FBGs, Part I: Elementary specimen. Composites Part A: Applied Science and Manufacturing **40**(1), 94–104 (2009). doi:10.1016/j.compositesa.2008.10.008
20. Olivier, P., Cavarero, M.: Comparison between longitudinal tensile characteristics of thin and thick thermoset composite laminates: influence of curing conditions. Computers & Structures **76**(1-3), 125–137 (2000). doi:10.1016/S0045-7949(99)00161-3

21. Pagano, N.J., Hahn, H.T.: Evaluation of Composite Curing Stresses. In: G.M.L. Gladwell, J.N. Reddy (eds.) Mechanics of Composite Materials, *Solid mechanics and its applications*, vol. 34, pp. 57–69. Springer Netherlands, Dordrecht (1994). URL `https://doi.org/10.1007/978-94-017-2233-9_6`
22. Prussak, R., Stefaniak, D., Hühne, C., Sinapius, M.: Evaluation of residual stress development in FRP-metal hybrids using fiber Bragg grating sensors. Production Engineering **12**(2), 259–267 (2018). doi:10.1007/s11740-018-0793-4
23. Prussak, R., Stefaniak, D., Kappel, E., Hühne, C., Sinapius, M.: Smart cure cycles for fiber metal laminates using embedded fiber Bragg grating sensors. Composite Structures **213**, 252–260 (2019). doi:10.1016/j.compstruct.2019.01.079
24. Schlimbach, J., Ogale, A.: Out-of-autoclave curing process in polymer matrix composites. In: Manufacturing Techniques for Polymer Matrix Composites (PMCs), pp. 435–480. Elsevier (2012). doi:10.1533/9780857096258.3.435
25. Shimizu, T., Kotlik, J.C., Arafath, A.R.A., Poursartip, A.: Evaluation of temperature profiles in thick composite parts during autoclave processing. In: A.C. Loos, M.W. Hyer, American Society for Composites (eds.) ASC Series on Advances in Composite Materials: Manufacturing of Composites. Destech Publications Incorporated (2013). URL `https://books.google.de/books?id=LjN7Vi5Mt4AC`
26. Slesinger, N., Shimizu, T., Arafath, A.R.A., Poursartip, A.: Heat transfer coefficient distribution inside an autoclave. In: 17th International Conference On Composite Materials (2009)
27. Tsukada, T., Takeda, S., Minakuchi, S., Iwahori, Y., Takeda, N.: Evaluation of the influence of cooling rate on residual strain development in unidirectional carbon fibre/polyphenylenesulfide laminates using embedded fibre Bragg grating sensors. Journal of Composite Materials **51**(13), 1849–1859 (2017). doi:10.1177/0021998316662327
28. Tuttle, M.E., Brinson, H.F.: Resistance-foil strain-gage technology as applied to composite materials. Experimental Mechanics **24**(1), 54–65 (1984). doi:10.1007/BF02323207
29. Twigg, G., Poursartip, A., Fernlund, G.: An experimental method for quantifying tool–part shear interaction during composites processing. Composites Science and Technology **63**(13), 1985–2002 (2003). doi:10.1016/S0266-3538(03)00172-6
30. Wang, Q., Wang, L., Zhu, W., Xu, Q., Ke, Y.: Numerical investigation of the effect of thermal gradients on curing performance of autoclaved laminates. Journal of Composite Materials **54**(1), 127–138 (2020). doi:10.1177/0021998319859061
31. Wiedemann, J., Pittsyk, T., Völkerink, O., Hühne, C.: Temperature distribution inside composite and fiber metal laminates during modified cure cycles. Procedia CIRP **131**, 107–112 (2025). doi:10.1016/j.procir.2024.10.333
32. Wiedemann, J., Prussak, R., Kappel, E., Hühne, C.: In-situ quantification of manufacturing-induced strains in fiber metal laminates with strain gages. Composite Structures **Vol. 691**(1), 115967 (2022). doi:10.1016/j.compstruct.2022.115967
33. Wisnom, M.R., Gigliotti, M., Ersoy, N., Campbell, M., Potter, K.D.: Mechanisms generating residual stresses and distortion during manufacture of polymer–matrix composite structures. Composites Part A: Applied Science and Manufacturing **37**(4), 522–529 (2006). doi:10.1016/j.compositesa.2005.05.019
34. Zappino, E., Petrolo, M., Santori, M.: Layer-Wise Modeling of Temperature Distributions and Degree of Cure to Evaluate Process-Induced Deformation and Residual Stress. In: ASME 2024 Aerospace Structures, Structural Dynamics, and Materials Conference. American Society of Mechanical Engineers (2024). doi:10.1115/SSDM2024-121553
35. Zappino, E., Santori, M., Masia, R., Zobeiry, N., Petrolo, M.: Numerical Analysis of the Impact of Process Parameters on the Residual Stress of a Flat Composite Part. Aerotecnica Missili & Spazio (2024). doi:10.1007/s42496-024-00231-7
36. Zobeiry, N., Poursartip, A.: The origins of residual stress and its evaluation in composite materials. In: P.W.R. Beaumont (ed.) Structural integrity and durability of advanced composites, Woodhead publishing series in composites science and engineering, pp. 43–72. Elsevier, Amsterdam (2015). doi:10.1016/B978-0-08-100137-0.00003-1

Chapter 5
Residual Stress Quantification Using Asymmetric Laminates

Abstract The curvature and stress-free temperature derived from asymmetric specimens after manufacturing can be used to quantify residual stresses in FML laminates. Several intrinsic and extrinsic parameters influence the accuracy of this experimental method. To determine the residual stress state in one principal direction, specimens with a small width-to-length ratio are preferred. Using three-dimensional curvature or stress-free temperature measurements in a lab oven provides accurate process-specific parameters from which the residual stress state can be derived. The accuracy is in the same order of magnitude compared to measurements using integrated sensors. Since the stress-free temperature is a parameter that is independent of the specimen layup, it can be directly transferred to laminates with other layups, including symmetrical ones. The only restriction is that the extrinsic parameters influencing the residual stress development in asymmetric specimens are not scaling with specimen size and that the manufacturing conditions (e.g. prepreg material, tool material, vacuum bagging setup, temperature profile of the cure cycle) are the same for both layups.[1]

5.1 Motivation

Measuring residual stresses using sensors discussed in the previous chapter is challenging outside laboratory environments. Integrating sensors into industrial manufacturing processes is complex and resource-intensive. As a solution, this chapter investigates an alternative stress quantification method that does not need any integrated sensors but uses specimens with an asymmetric layup. Using these specimens is an excellent option for industrial applications since the complexity during manufacturing is low (compared to instrumented specimens), and the evaluation can be executed post-manufacturing.

[1] Parts of the results of this chapter have been published in [23]. Minor deviations of some parameters from earlier publications result from differences in the material properties assumed in these studies.

J. Wiedemann, *Relevance of Manufacturing-Induced Residual Stress States in Fiber Metal Laminates for Structural Health Monitoring*, Mechanics and Adaptronics, https://doi.org/10.1007/978-3-032-09925-9_5

Fig. 5.1 FML specimens with an asymmetric layup $[St/0°_7]$ manufactured on a flat tool curve after demolding at room temperature.

The idea behind asymmetric layups is the effect of a bi-material that starts to curve when subjected to a temperature change due to the different CTEs of the two materials. In composite manufacturing, the bi-material can consist of plies with two different fiber orientations or metal and FRP in the case of FMLs. The asymmetric layup is placed on a flat tool, and due to the bonding of the single layers at elevated temperatures, the specimens start to produce a curvature after demolding. This is illustrated in the photograph in Fig. 5.1. The curvature is a measure of the residual stress that develops in the laminates during the manufacturing process.

The use of asymmetric specimens is very prominent in the literature [4, 5, 21] to determine the residual stresses developing in thermoplastic and thermoset CFRPs. Beyond these investigations, others used the curvature analysis for asymmetric FML specimens [14, 15] or to investigate the stress influence of an adhesive layer on a steel sheet [25]. However, Zobeiry and Poursartip [26] state that warpage or curvature measurements go along with limited accuracy of the results. Likewise, in [14], different evaluated parameters of curved laminates (curvature and stress-free temperature) show large deviations among themselves. It is assumed that the layup, laminate thickness, and dimensions of the specimens play a significant role in the evaluation reliability, in addition to the manufacturing process.

To get an overview of specimen dimensions and layups used in the literature, Table 5.1 lists some sources with their corresponding asymmetric specimen architectures made of monolithic CFRPs or FMLs. The size of the specimens and the layups vary significantly. For small aspect ratios (width-to-length), the curvature in the transverse direction only influences the primary curvature to a minor degree. With aspect ratios close to 1, a saddle-shape curvature or a bi-stable deflection behavior is expected [4, 8].

Asymmetric laminates can be analyzed either in terms of their curvature or stress-free temperature [11, 12, 16]. While the curvature can be directly measured, the stress-free temperature is determined as the temperature at which the curved laminates transition back into their plane state when the specimens are reheated. Most of the time, the stress-free temperature is somewhere below the final cure temperature of the laminate [3, 18].

Table 5.1: Asymmetric specimens made of CFRPs or FMLs used in the literature vary greatly in their dimensions and laminate layups.

Material	Source	Dimensions in mm	Width-to-length ratio	Layup	Thickness in mm
CFRP	[17]	25 x 152	0.16	$0_3/90_3$	0.75
CFRP	[21]	100 x 230[a]	0.4	$0/90$	0.25
CFRP	[5]	25/11 x 152	0.16/0.07	$0_4/90_4$	-
CFRP	[4]	25 x 150	0.16	$90_4/0_2, 90_2/0_2, 90_2/0$	0.4-0.8
FML	[14]	20 x 200	0.1	$St/0$	0.25
FML	[15]	100 x 200	0.5	$St/0_3$	0.5

[a] trimmed

Both parameters, curvature, and stress-free temperature, are sensitive to several variables contributing to the residual stress state. These variables can generally be divided into intrinsic and extrinsic parameters [1,9]. Intrinsic parameters include material and layup-dependent anisotropy, whereas extrinsic parameters include tool and process influences. The tool is a main extrinsic parameter affecting the residual stress state, which mainly results from incompatibility in thermal expansion between the tool and laminate. Two phenomena are differentiated in the literature: forced interaction [9,13] and warpage [9,19]. Since the laminates are manufactured in a plane state, no geometrical interlocking between the tool and composite part is present. Therefore, only the effect of warpage needs to be considered. Warpage mainly occurs for thin laminate thicknesses ($t \leq 1$ mm) in combination with significant CTE differences between tool material and laminate [1]. Table 5.1 shows that the specimen architectures of the asymmetric specimens used in the literature are within the relevant thickness range for warpage. Hence, this effect needs to be considered during the evaluation.

The range of influences shows that a comprehensive understanding of the effect of the different parameters on the residual stress state and, consequently, on the curvature of an asymmetric specimen is necessary. To develop a reliable method that uses curvature analysis of asymmetric specimens for process monitoring and residual stress quantification in more complex FML structures, the following hypothesis is formulated:

! Research hypothesis 2

Asymmetric specimens can determine the residual stress states in FMLs with any dimension and layer structure by deriving a parameter only sensitive to extrinsic manufacturing process variables.

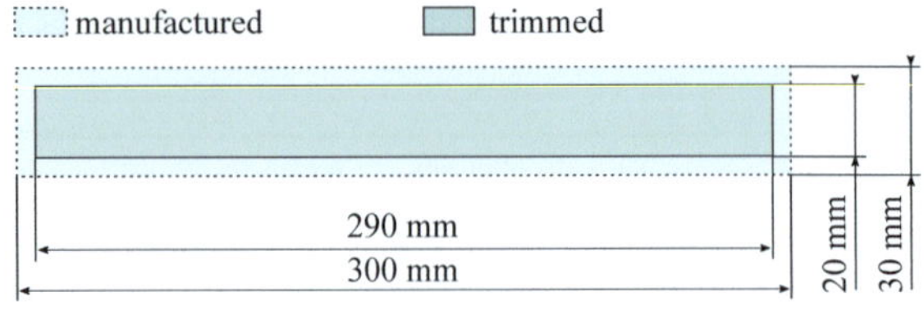

Fig. 5.2 Dimensions of the as manufactured and trimmed specimens in their plane state (adapted from [23]).

5.2 Experimental Setup and Specimen Program

During the experiments, a specimen size with a width-to-length ratio of 0.07 is used. This way, the specimen curvature will be dominant in one direction, and the influence of the transverse curvature can be minimized. All specimens are manufactured using the two manufacturing processes (MRCC and MOD) introduced in Chap. 3. Although the tool material varied during the investigations, the tool and cover plate materials were always identical for all specimens.

The specimens are manufactured with a size of 300 mm by 30 mm with a subsequent trimming step to reduce the influence of edge effects. These effects originate from a slight decrease in thickness towards the specimen edges due to resin bleed. From the GOM ATOS measurement (Fig. 5.3), it can be seen that the thickness of the trimmed specimen is very much evenly distributed over the entire specimen. Figure 5.2 illustrates the dimensions during manufacturing and after trimming. It must be emphasized that manufacturing and trimming are always performed in the specimens' flat state by applying external loads. The curvature of the specimens is restored when the external forces are relieved.

The trimming process can influence the curvature of the specimen. It was found that trimming using a circular saw with a diamond blade leads to uneven edges and local edge delamination in the specimens. Additionally, the process may result in considerable heat generation that can change the specimen state. Therefore, water jet cutting was used to trim the specimens. This method guarantees high cutting accuracy. It also prevents heat generation in the specimens, significantly reduces the roughness of the edges, and eliminates any edge delamination.

To account for many influencing variables on such asymmetric specimens, the influence of two intrinsic and three extrinsic parameters on the asymmetric specimens' curvature and stress-free temperature is investigated. The specimen layup and the CFRP prepreg batch are the two intrinsic parameters, the tool material, the cure cycle, and the manufacturing setup are the three extrinsic parameters considered. Table 5.2 gives an overview of all investigated parameters.

To investigate the influence of these parameters, a specimen program consisting of 47 asymmetric laminates is manufactured. Table 5.3 gives an overview of all the specimens and their respective parameters. Every specimen is given an ID during the manufacturing process. Subsequent representations condense specimens with identical influencing parameters into a single ID. The mean value and the respective standard deviation are always shown when the results are presented.

Table 5.2: Parameters influencing the residual stress state and their variations investigated in this work.

Parameter	Variation
Layup	No. of CFRP and steel plies (varying laminate thicknesses and metal volume fractions)
Prepreg batch	Different initial degrees of cure
Manufacturing setup	Single specimen on tool and two specimens stacked on top of each other
Tool material	Steel (S) and aluminum (A)
Cure cycle	Manufacturer recommended (MRCC) and modified cure cycle (MOD)

Table 5.3: Overview of the specimen program and the parameter variations.

	Intrinsic parameters:		Extrinsic parameters:			
ID	Layup	CFRP batch	Tool	Cure cycle	Setup	Quantity
13AR-e	$St/0_3$	expired	A	MRCC	single	3
14SR	$St/0_4$	new	S	MRCC	single	4
14SR-s	$St/0_4$	new	S	MRCC	stack	4
14SM	$St/0_4$	new	S	MOD	single	2
14AR	$St/0_4$	new	A	MRCC	single	1
14AR-e	$St/0_4$	expired	A	MRCC	single	3
14AR-s	$St/0_4$	new	A	MRCC	stack	2
14AM	$St/0_4$	new	A	MOD	single	2
15AR	$St/0_5$	new	A	MRCC	single	3
17SR	$St/0_7$	new	S	MRCC	single	4
17SR-s	$St/0_7$	new	S	MRCC	stack	4
17SM	$St/0_7$	new	S	MOD	single	3
17AR	$St/0_7$	new	A	MRCC	single	1
17AR-e	$St/0_7$	expired	A	MRCC	single	3
17AR-s	$St/0_7$	new	A	MRCC	stack	3
17AM	$St/0_7$	new	A	MOD	single	3
1412AR	$St/0_4/St/0_2$	new	A	MRCC	single	2
					SUM	47

The sample identifier reflects the variations in the parameter space. An explanation of the specimen labeling is given in Table 5.4 using the specimen with ID14SR-s as an example. The first two digits indicate the number of steel and CFRP plies within the laminate. The following two digits specify the tool material the specimen was manufactured on and the cure cycle. The last digit indicates the manufacturing setup, and the CFRP batch used.

Table 5.4: Explanation of the logic used to label the manufactured specimens with specimen ID14SR-s used as an example.

1	No. of steel plies	-
4	No. of CFRP plies	-
S	Tool material	S: steel / A: aluminum
R	Cure cycle	R: recommended / M: modified
-s	Additional information	-s: stacked manufacturing setup / -e: expired CFRP batch

5.3 Experimental Methods for Parameter Determination From Asymmetric Specimens

5.3.1 Curvature Evaluation

Different methods exist to evaluate the curvature of such asymmetric laminates after manufacturing. The simplest method is to measure the width and height of the specimens when placed on a flat surface. Hence, the curvature can be calculated assuming a uniform deformation [15]. Alternatively, a two-dimensional contour plot of such a specimen can be determined with a laser distance sensor attached to a two-dimensional gantry. This technique evaluates the distance between the curved specimen and sensor (Fig. 5.3a), and consequently, the specimen's radius is determined by calculating a fitting curve. However, this approach proved very time-consuming and included difficulties for large curvatures due to the measurement range of the laser sensor and reflection issues. Therefore, an industrial three-dimensional scanning head (GOM ATOS) is used for the subsequent experiments. The GOM ATOS scan generates a point cloud of the specimen's surface, which is analyzed using the GOM-Inspect software. By fitting a cylinder into the point cloud, the radius of the specimens is determined (see Fig. 5.3b). Another advantage of using the GOM ATOS scan is the possibility of evaluating the laminate thickness and the thickness distribution.

5.3.2 Stress-Free Temperature Evaluation

The stress-free temperature of all specimens is determined in a lab oven with a glass window. Figure 5.4 shows the setup for a single specimen. The specimens are placed on a metal plate in the oven. They are subsequently heated from room temperature to above 100 °C depending on the expected stress-free temperature with a heating ramp of $4\,\mathrm{K\,min^{-1}}$. At this temperature, they are allowed to equalize for 30 min before being heated slowly with a heating rate of $0.7\,\mathrm{K\,min^{-1}}$ up to the stress-free temperature. The stress-free state is visually determined when all specimen edges are in contact with the bottom plate (see Fig. 5.4b). The temperature is measured with several type K thermocouples on the bottom plate to ensure that the

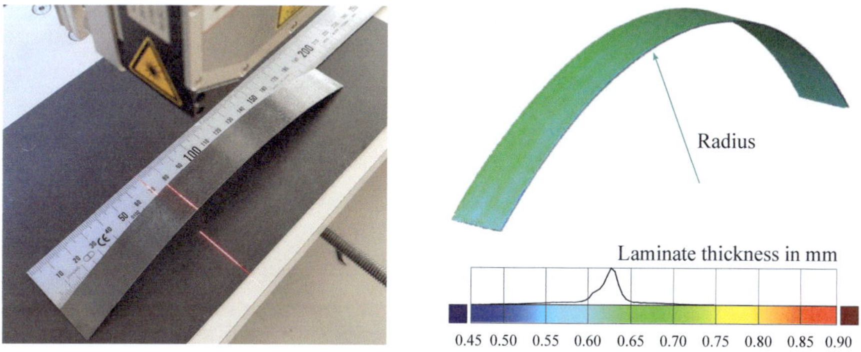

(a) (b)

Fig. 5.3: Two different methods to experimentally evaluate the curvature of asymmetric FML specimens: **(a)** two-dimensional contour measurement with a laser distance sensor and **(b)** three-dimensional curvature measurement with the GOM ATOS system (adapted from [23]).

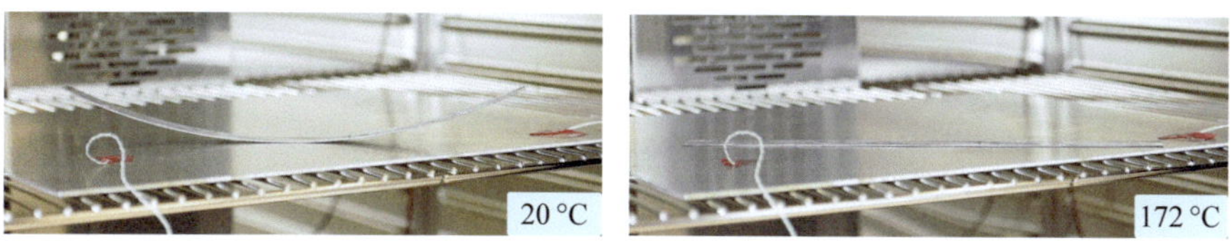

(a) (b)

Fig. 5.4: The curvature of an asymmetric FML specimen at room temperature **(a)** decreases with increasing temperature, and **(b)** the specimen reaches a flat state at its stress-free temperature (adapted from [23]).

temperature is distributed homogeneously. Determining the stress-free temperature is a visual technique and highly depends on the person experimenting. Therefore, the accuracy of the measurements is analyzed by repeating the measurement for identical specimens. The resulting accuracy of this technique is within a range of ±3 °C.

5.4 Results of the Experimental Curvature Analysis

Every single specimen is evaluated for its respective curvature and stress-free temperature[2]. The following results will be discussed using the curvature measured with the 3D GOM ATOS system. Nevertheless, the same conclusions can be drawn using

[2] The experiments were carried out in cooperation with Jan-Uwe R. Schmidt during his time as a research assistant at the Institute of Mechanics and Adaptronics of the Technische Universität Braunschweig.

Table 5.5: Layup variations investigated in this work and resulting MVFs and nominal laminate thicknesses. Layups in bold print are chosen for the subsequent investigations.

ID	Layup	Metal volume fraction in %	Nominal thickness in mm
13...	$St/0_3$	24	0.51
14...	$\boldsymbol{St/0_4}$	**19**	**0.64**
15...	$St/0_5$	16	0.77
17...	$\boldsymbol{St/0_7}$	**12**	**1.03**
1412...	$St/0_4/St/0_2$	24	1.02

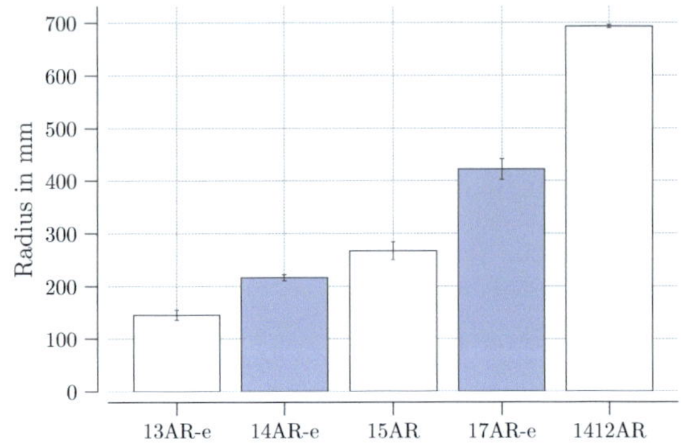

Fig. 5.5 Variations in the layup and consequent specimen thicknesses result in significantly different radii of the asymmetric specimens. The layups with ID14 and ID17 (blue color) are selected for further analysis.

the stress-free temperature since a direct correlation between the two parameters exists, which will be explained in Sec. 5.4.3.

To avoid confusion during the presentation of the results, the intrinsic parameters are evaluated independently of the extrinsic parameters for a specific extrinsic parameter configuration (aluminum, MRCC, single). In contrast, the extrinsic parameters are evaluated for two different layups and the same CFRP prepreg batch.

5.4.1 Intrinsic Parameter Influences

Laminate Layup

The literature presents many layup configurations for asymmetric laminates (cf. Table 5.1). Five layup configurations that differ in ply stacking, metal volume fraction, and thickness are compared to evaluate the suitability of certain layups. Table 5.5 shows the corresponding layups and values.

Figure 5.5 shows the curvature evaluation of these specimens. It can be seen that with increasing specimen thickness, except for layup ID1412, the radius of the respective specimens increases. Specimens with layup ID13 produced minimal radii. Due to their small laminate thickness and consequent low bending stiffness, the curvature measurement can be influenced by gravitation, which depends on the

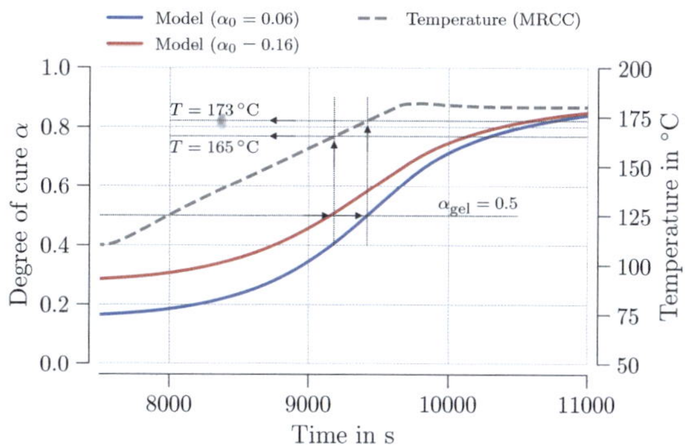

Fig. 5.6 An increase of the prepregs initial degree of cure from $\alpha_0 = 0.06$ to $\alpha_0 = 0.16$ results in the gel point (assumption: $\alpha_{gel} = 0.5$) being reached significantly earlier in time and temperature during the MRCC based on the cure kinetics model.

placement of the specimen during the measurement. Specimens with ID1412 are no longer covered by the analytical solution (shown in a later section) due to their complex layup. Hence, the two laminates, ID13 and ID1412, are excluded from further analysis. Likewise, the layup ID15 is excluded to have a greater difference in thickness and MVF between the remaining layups. The layups with ID14 and ID17 are consequently selected for all the subsequent investigations to determine the influence of the other intrinsic and extrinsic variables on the curvature.

Initial Cure of Prepreg Material

During the investigations, two different batches of CFRP material were used to assess the influence of the initial degree of cure of the prepreg material. The first prepreg batch's shelf life (12 months from the date of manufacture [7]) has already expired, while the second batch has just been delivered. Due to many freeze-thaw cycles and the amount of out-time (days out of the freezer), the initial degree of cure of the expired material differs from the new material. Grunenfelder et al. [6] showed that during 14 days of out-time, an epoxy prepreg material's initial degree of cure increases from $\alpha_0 = 0.055$ to $\alpha_0 = 0.148$. It can be assumed that the expired material has been stored at room temperature for at least two weeks. Using the cure kinetics model presented earlier (cf. Sec. 3.5), Fig. 5.6 shows that an increase of 10% in initial cure leads to the gel point being reached 8 K earlier during the MRCC. Consequently, it can be assumed that the bonding temperature is likewise reached earlier during the process, leading to reduced residual stresses or, in terms of curvature, an increase in specimen radius.

Figure 5.7 shows that the curvature evaluation of the specimens manufactured from the expired CFRP prepreg batch results in larger radii than those manufactured with the new material for otherwise identical layups and manufacturing conditions. Therefore, the initial state of the prepreg material also needs to be kept in mind when calculating residual stresses based on handbook formulas.

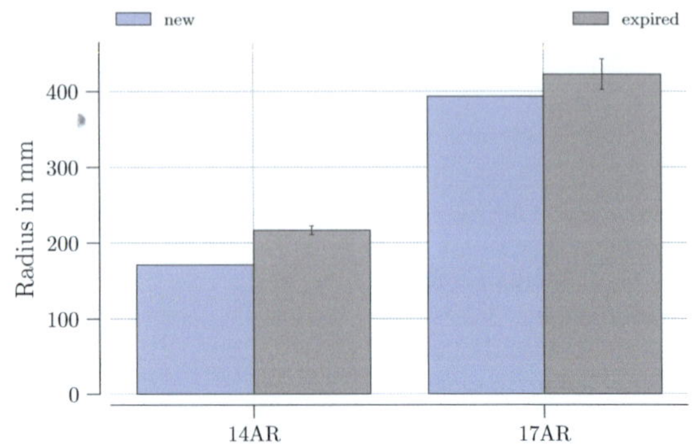

Fig. 5.7 Specimens manufactured with the expired CFRP prepreg batch produce larger radii than those manufactured with the new prepreg material.

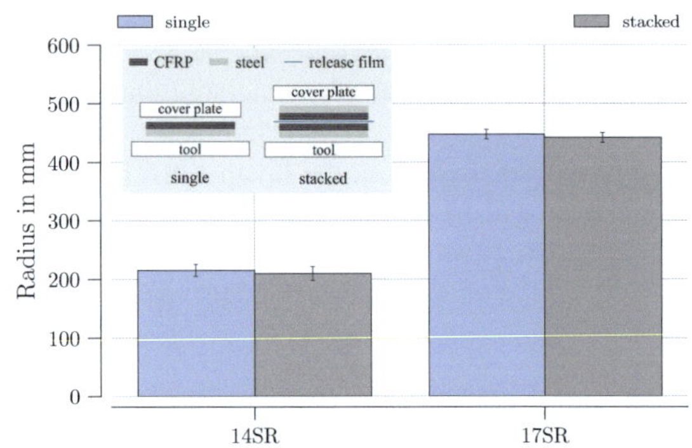

Fig. 5.8 The manufacturing setup does not significantly impact the curvature of the specimens.

5.4.2 Extrinsic Parameter Influences

Manufacturing Setup

Different setups for manufacturing asymmetric specimens can be found in the literature. While most researchers manufactured every asymmetric laminate by itself, [14] used a configuration in which two asymmetric specimens were stacked on top of each other and separated by a thin release film. The release film allows the separation of the two specimens after manufacturing. The advantage of this setup is less space requirement on the tool surface and the need for only one cover plate per two specimens.

Figure 5.8 illustrates the difference between the two setups and the results of the curvature measurements. The results show that the manufacturing setup does not show any significant influence on the specimen's curvature. However, visual inspection of the specimens manufactured with the stacked setup revealed that the CFRP surface of these specimens shows a high roughness. This roughness of the stacked specimens can be explained by the lack of a rigid and smooth metal tool or cover plate on the CFRP surface. This leads to the fibers interacting with the surface of the opposite specimen since they are only separated by a thin release

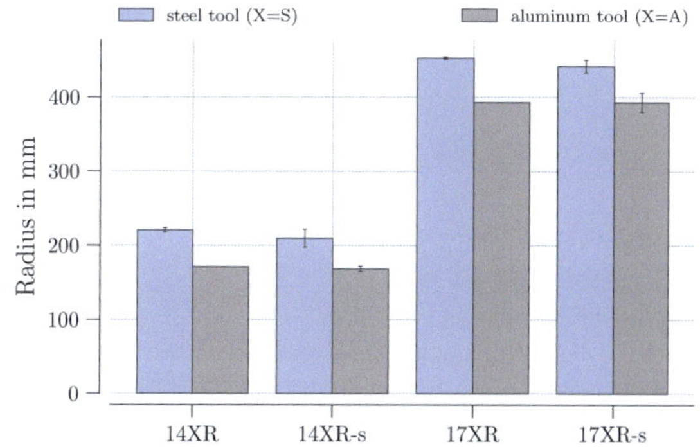

Fig. 5.9 Specimens manufactured on the aluminum tool show significantly smaller radii than those manufactured on the steel tool.

film. This effect might increase the variation during curvature evaluation but was not observed here. It is assumed that reversing the order so that the two steel surfaces are separated by the release film will eliminate this effect. For the single setup, no influence was found for reversing the order of CFRP and steel, and hence, the stacked setup with reversed order seems an appropriate option for manufacturing asymmetric specimens.

Tool Material

One great extrinsic contributor to the residual stress level in an FML is the material of the tool [24]. Therefore, two common tool materials in composite manufacturing, steel and aluminum, are chosen to compare their influence on the specimen's curvature. The results are shown in Fig. 5.9 for the two selected layups (ID14 and ID17).

Specimens manufactured on the aluminum tool show a smaller radius for all investigated configurations than those manufactured on the steel tool. This can be attributed to the mismatch of the coefficients of thermal expansion between laminate and tool, which results in warpage of the specimens after manufacturing [19]. The steel tool and steel cover plates show a similar thermal expansion behavior as the steel plies in the laminate themselves (approx. $19 \times 10^{-6}\,K^{-1}$, cf. Table 3.2). However, the aluminum tool and cover plates possess higher coefficients of thermal expansion of around $24 \times 10^{-6}\,K^{-1}$. Hence, when using the aluminum tool, the layers close to the tool surface are pre-strained during the heating phase before curing. This results in a strain gradient through the thickness of the specimens. These strains are superposed with the thermal cure strains resulting from the thermal incompatibility between steel and CFRP layers and are locked into the laminate during resin solidification. After demolding, this results in decreased radii than for the specimens manufactured on the steel tool. Although the specimens are separated from the tool using release film, the autoclave pressure and vacuum increase the friction coefficient between the laminate and the tool. To reduce residual stress levels, it is recommended to use tool materials with CTEs similar to those of the laminate materials.

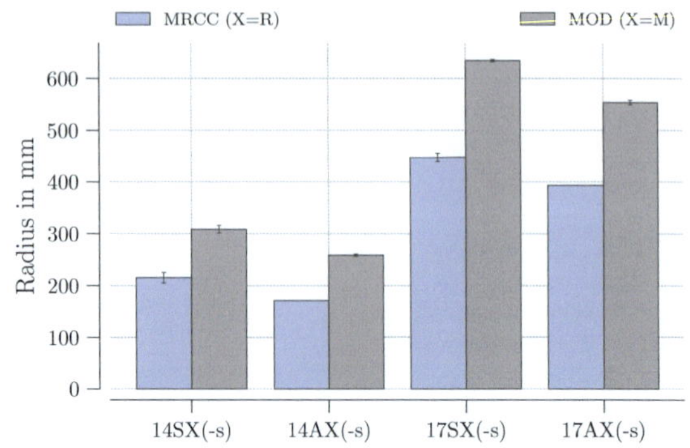

Fig. 5.10 The modified cure cycle (MOD) increases the radius of asymmetric specimens compared to the manufacturer-recommended cure cycle (MRCC), independent of tool material and specimen layup.

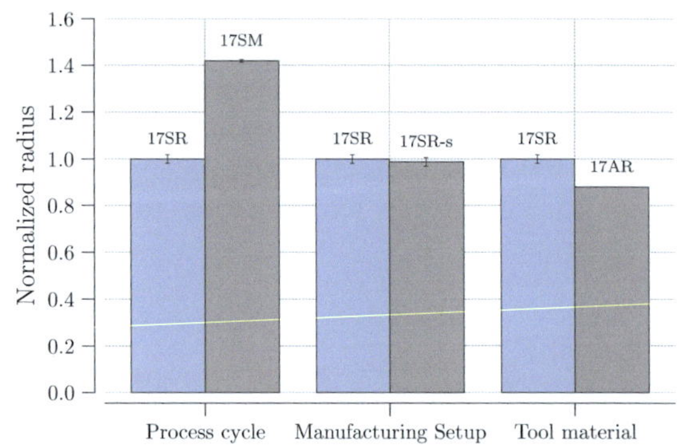

Fig. 5.11 The cure cycle is the extrinsic parameter that contributes the most to the specimen curvature compared to the other two extrinsic parameters: tool material and manufacturing setup. ID17SR is used as the reference curvature.

Cure Cycle

Two different autoclave cure cycles (MRCC and MOD) and their influence on specimen curvature are investigated. Figure 5.10 shows the resulting curvatures for specimens ID14 and ID17 for the two different tool materials. It is obvious that the modified cure cycle results in significantly larger radii than the MRCC independent of other intrinsic and extrinsic parameters. For specimens with ID14 the radius is increased by 42 to 50% and for specimens with ID17 by around 40%.

Figure 5.11 compares the relative influence of all three different extrinsic parameters investigated. It becomes clear that the most significant change in curvature results from modifying the cure cycle, followed by the tool material, while the manufacturing setup can be considered negligible.

5.4.3 Correlation of Curvature and Stress-Free Temperature

Several analytical models to describe the curvature of asymmetric laminates after manufacturing are proposed in the literature [2, 10, 20]. The analytical approaches

Fig. 5.12 Numerical model of a specimen with layup ID17 used for the reverse calculation to get the deformation for a given change in temperature.

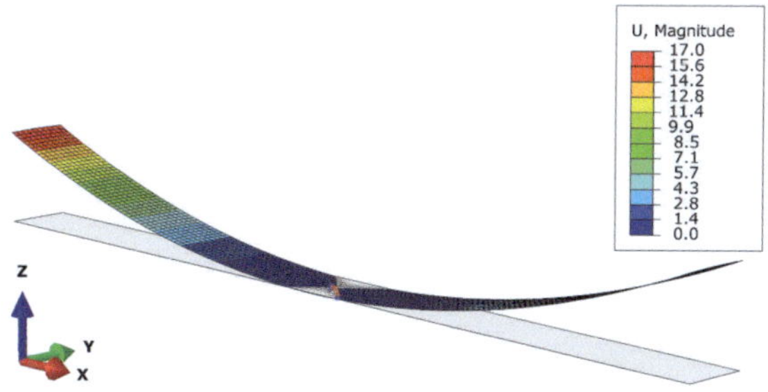

mainly originate from bimetal applications with two different metallic layers combined to produce a thermomechanical bending transducer. The most comprehensive analytical description to calculate the radius r of such bi-materials dependent on a reference temperature is given by Timoshenko [20]:

$$r = \frac{h(3(1+m)^2 + (1+m\,n)(m^2 + \frac{1}{m\,n}))}{6(\alpha_2 - \alpha_1)(T_{\mathrm{sf}} - T_{\mathrm{r}})(1+m)^2}\text{, with} \tag{5.1}$$

$$m = \frac{t_1}{t_2}\text{, and } n = \frac{E_1}{E_2} \tag{5.2}$$

where h is the specimen thickness, α_1 and α_2 are the coefficients of thermal expansion for the two materials, T_{r} is the reference temperature, and T_{sf} is the stress-free temperature. $m = t_1/t_2$ represents the ratio of the thicknesses (t_1 and t_2) of the two constituents (FRP and steel), whereas $n = E_1/E_2$ indicates the ratio of the elastic moduli of the two constituents (FRP and steel), respectively.

This equation is widely used in the literature to describe curvatures of asymmetric CFRP laminates [4,10] and FML [16] and therefore also used in this work to correlate the radius and stress-free temperature of different asymmetric laminates with each other.

Additionally, numerical simulations are used to validate the analytical curvature calculation. In the finite element analysis (FEA), the experiment to determine the stress-free temperature based on a given curvature is modeled using the software Abaqus. A curved specimen is created using a shell section and given the radius derived from the GOM ATOS measurements. The laminate layup and the respective material properties are assigned to the shell section, and the part is meshed using quadrilateral shell elements with reduced integration (type S8R). Subsequently, the specimen is placed on a flat surface of rigid body elements. Because of the large deformations, a non-linear calculation is pursued. A temperature step is included in the model to achieve the deformation by heating the specimen from room temperature to a higher temperature level. Using a Python script, the temperature at which the curved specimen transitions into its flat state is determined iteratively. The ideal flat state is defined when the nodes at the short specimen edges have identical Y-coordinates compared to the center nodes, as shown in Fig. 5.12. Consequently, the

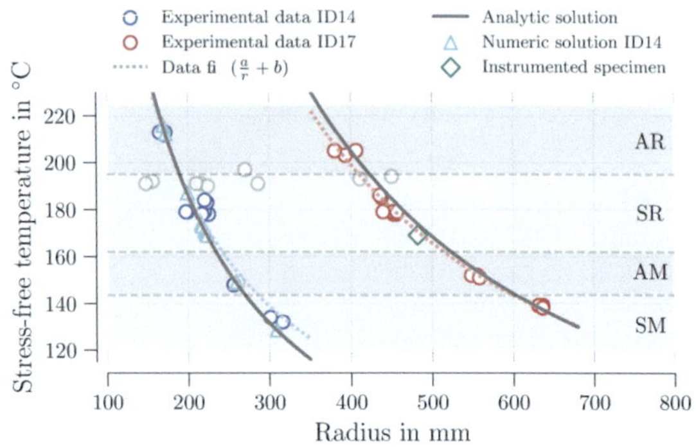

Fig. 5.13 Correlation of stress-free temperature with specimen radius and comparison to analytical and numerical derived values. Specimens can be clustered according to their stress-free temperature and the two extrinsic influencing variables, tool material, and cure cycle.

Table 5.6: Experimental T_{sf} for the 8552-AS4 prepreg material for different extrinsic parameter changes (based on asymmetrical specimens with ID14 and ID17 shown in Fig. 5.13).

	AR	SR	AM	SM	
T_{sf}	209 ± 5	180 ± 3	150 ± 2	136 ± 3	°C

resulting temperature is defined as the stress-free temperature of the investigated specimen.

Figure 5.13 shows the experimentally derived values for curvature and stress-free temperatures for all specimens. The focus is on the specimens with layups ID14 and ID17. All other layup variations (ID15, ID1412) and specimens manufactured with the expired CFRP batch are indicated by the light grey markers but eliminated from further analysis. A curve in the form $a/r + b$ is fitted through the experimental data, with the radius r and the fitting parameters a and b. The fit and the analytic solution are in good agreement for both layups. For the layup ID14, the numerically calculated stress-free temperatures for specific radii are calculated and added to the figure. The numerical simulation results are congruent with the analytical solution. The same applies to specimens with layup ID17, although the figure does not show the numerical data points.

The different layups, ID14 and ID17, can be clearly distinguished by their radii. However, the stress-free temperatures are relatively independent of the layup and only sensitive to the extrinsic parameters: tool material (A: aluminum and S: steel) and cure cycle (R: MRCC and M: MOD). This is indicated by the horizontal shaded areas in Fig. 5.13. Table 5.6 provides the mean stress-free temperatures for these extrinsic parameter combinations. If one compares the values with the literature suggestions for calculating the stress-free temperature, given in Table 2.2, it becomes clear that the deviations between the literature suggestions can all be achieved by only changing the extrinsic process parameters. This also shows that calculating the stress-free temperature based on handbook formulas is inaccurate without considering the manufacturing setup and the cure cycle.

Table 5.7: Experimentally determined parameters with the instrumented specimen compared to analytically calculated values at a reference temperature of $T_r = 23$ °C. The strain delta indicates the difference between measured and calculated strains.

	Radius in mm	T_{sf} in °C	Strain (flat) in µm m^{-1}	Strain delta in %
Experiment	GOM ATOS: 482	Oven: 169	Strain gauge: 2189	-
Analytic	497	169	→ 2289	+4.6%
Analytic	482	173	→ 2351	+7.4%

5.4.4 Validation Using the Strain Gauge Method

One asymmetric specimen with layup ID17 is equipped with a strain gauge according to the approach discussed in the previous chapter to validate the applicability of asymmetric specimens for the residual stress quantification in FML. The specimen is placed with the CFRP layers onto the tool surface. The strain gauge bonded to the top metal ply is contacted through a cut-out in the cover plate. The dummy strain gauge bonded to an ultra-low expansion (ULE) glass specimen is placed next to the asymmetric specimen in the autoclave to compensate for any temperature-induced strains in the strain gauge itself. The temperatures are monitored with thermocouples placed inside the laminate and on the ULE glass specimen.

The resulting curvature and stress-free temperature of that instrumented specimen are also shown in Fig. 5.13. It correlates very well with the analytical solution. However, it slightly deviates from the other specimens manufactured with the same tool and cure cycle. There are a few reasons for this deviation. One reason might be slight differences in the manufacturing process. A second reason might be the increase in specimen stiffness due to the attached strain gauge furthest away from the specimen midplane. Finally, the edges of the specimen were left untrimmed to avoid damaging the strain gauge, which might also have had an impact.

Table 5.7 compares the measured strains of the instrumented specimen with calculated strains using CLT and Eq. (5.1). The strain gauge measurements result in a final absolute strain of 2189 µm m^{-1} in the asymmetric specimen's flat state. With the experimentally determined stress-free temperature of that specimen, the CLT calculates a residual strain of 2289 µm m^{-1}. Using the measured radius and transforming it into a stress-free temperature with Eq. (5.1) and consequently calculating the residual strains with the CLT results in 2351 µm m^{-1}. Therefore, depending on the method, a relative error of 4.6% and 7.4% to the measured residual strains is made. Comparing this variation to the strain gauge measurements made with different strain gauges on identical specimens (Table 4.1: 1.6% to 7.0%), it can be concluded that no matter which parameter of the asymmetric specimen is evaluated, still a good agreement between curvature or stress-free temperature and measured strains is found.

Table 5.8 Residual tensile stresses in the metal sheets (primary direction) for each layup configuration and depending on the cure cycle.

	ID2	ID3a	ID4	ID6b	
MRCC	375	396	122	297	MPa
MOD	272	287	88	216	MPa
Delta	-103	-109	-34	-81	MPa

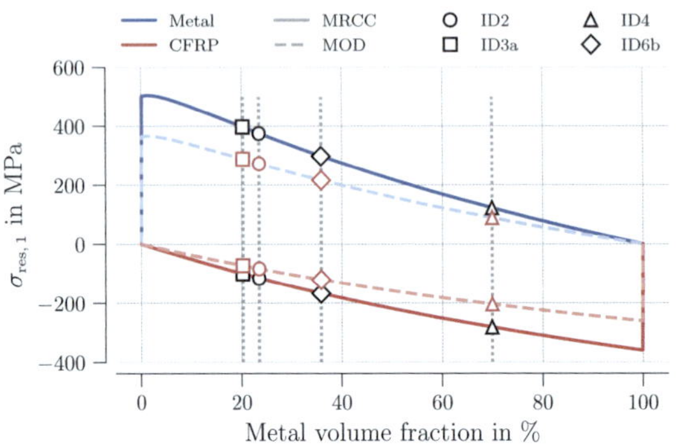

Fig. 5.14 Residual stresses (primary direction) in the metal and CFRP plies of an FML with uni-directional fiber orientation dependent on the metal volume fraction. The relevant laminate configurations of this work are indicated by their symbols (adapted from [22]).

5.5 Residual Stress Levels for Relevant Laminate Configurations

With the validated results of the asymmetric specimens for the two different cure cycles, the residual stress states for all the laminate configurations used in the following chapters (cf. Table 3.5) are presented in this section. All the relvant specimens are manufactured with the autoclave processes MRCC and MOD on a steel tool with a steel cover plate. Table 5.6 provides the relevant stress-free temperatures for the two relevant manufacturing conditions SR (MRCC) and SM (MOD). With these stress-free temperatures, the CLT, and the material properties from Chap. 3, the corresponding residual stress states for the two cure cycles can be calculated for each laminate configuration. Table 5.8 provides the corresponding residual tensile stresses in the FMLs in the primary direction of the metal sheets for each laminate and cure cycle.

Moreover, Fig. 5.14 shows the corresponding residual stresses in the primary direction of the FML laminates (rolling direction of the steel and fiber direction of the prepreg layers) as a function of the metal volume fraction. The stress states of the relevant laminates are also indicated with their corresponding symbol. From this figure, the residual stress states for any other laminate configurations can likewise be extracted.

Fig. 5.15 shows the residual stresses in the transverse direction for completeness. As previously discussed, these are much smaller than the residual stresses in the fiber direction of the uni-directional FML configurations. Therefore, the focus in the following investigations is on the effects of the residual stresses in the primary direction. In Chaps. 6 and 7, only the influences of different SHM-related parameters in the fiber direction of the uni-directional FMLs are investigated.

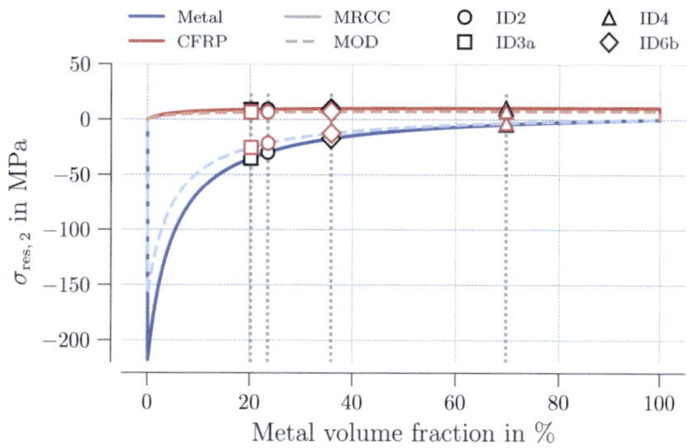

Fig. 5.15 Residual stresses (transverse direction) in the metal and CFRP plies of an FML with uni-directional fiber orientation dependent on the metal volume fraction. The relevant laminate configurations of this work are indicated by their symbols.

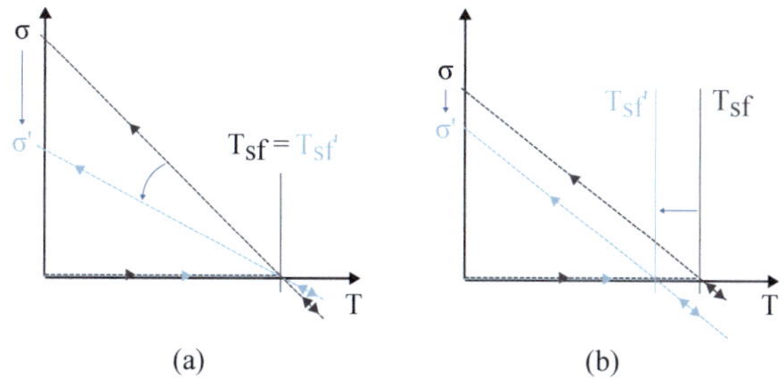

Fig. 5.16 Qualitative illustration of the difference in the residual stress evolution in the metal layers during the manufacturing process due to **(a)** a change of the layup and **(b)** a change in the cure cycle (considering linear thermo-elastic behavior and instantaneous bonding).

5.6 Conclusion and Discussion

The curvature and stress-free temperature of asymmetric specimens are well suited to capture the influence of extrinsic parameters, especially the cure cycle, on residual stress development. Both parameters, curvature and stress-free temperature, can be accurately determined using experimental methods. Curvature measurement using a three-dimensional scanning head is the preferred method since it can be completely automated. Stress-free temperature determination in an oven is also appropriate but requires more expertise.

The results confirm that asymmetric specimens reliably represent residual stresses in a laminate, provided certain aspects are considered. Particular attention must be paid to the design of the asymmetrical test specimens and the production setup. The results have shown that a small width-to-length ratio of the specimens seems favorable. Furthermore, the specimen thickness and tendency to be influenced by warpage due to thermal tool material incompatibility must be kept in mind. The influence of warpage is decreasing for thicker laminates, so derived residual stresses for thicker laminates might be less accurate. Therefore, the accuracy of the proposed method increases by keeping the influence of the tool within the asymmetric specimens to a minimum by using compliant tool CTEs.

Figure 5.13 has shown that the stress-free temperature is a process-specific parameter and independent of the layup of the asymmetric specimens. To further illustrate

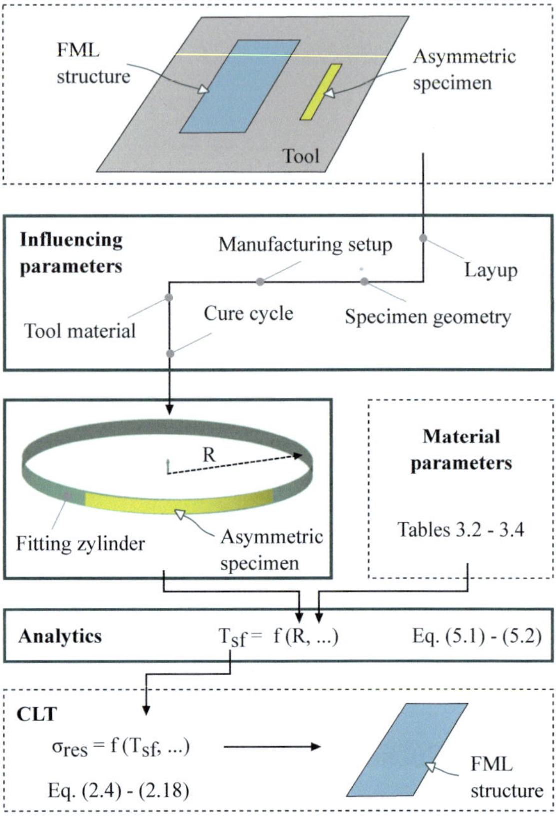

Fig. 5.17 Proposed approach to use asymmetric specimens for residual stress quantification of FML structures, including relevant equations and tables (adapted from [23]).

this finding, Fig. 5.16 qualitatively shows the difference between a change in the layup and the cure cycle on the residual stress evolution in the metal layers of an FML during the manufacturing process. Different layups lead to different CTEs of the laminate. However, this does not influence the process-specific T_{sf}. Contrastingly, a change in the cure cycle directly affects the temperature at which laminate behavior is established from the single layers and, with this, the T_{sf}. Consequently, the T_{sf} derived from the asymmetric specimens can be directly transferred to symmetric FMLs manufactured within the same manufacturing process on the same tool to quantify their residual stress states, although their layup (intrinsic parameter) differs from that of the asymmetric specimens.

It must be emphasized that the shown independence is only valid if extrinsic parameters do not scale with specimen size and manufacturing conditions for laminates of interest are identical (e.g. prepreg material, tool material, vacuum bagging setup, temperature profile of the cure cycle). Therefore, additional studies are necessary to unconditionally confirm this correlation for various symmetric layup configurations. Given these results, the hypothesis addressed in this chapter, *asymmetric specimens can determine the residual stress states in FMLs with any dimension and layer structure by deriving a parameter only sensitive to extrinsic manufacturing process variables*, can only be partially confirmed. The final approval needs to be obtained in future studies.

To ensure all relevant aspects are considered when asymmetric specimens are used for residual stress quantification, Fig. 5.17 suggests an approach to be followed. Asymmetric specimens accompanying more complex FML structures during manufacturing are evaluated in terms of their curvature. With the relevant material parameters, the curvature is transferred into a corresponding stress-free temperature. The T_{sf} is consequently used as an input for calculating the residual stress state within the complex FML structure by the CLT.

References

1. Albert, C., Fernlund, G.: Spring-in and warpage of angled composite laminates. Composites Science and Technology **62**(14), 1895–1912 (2002). doi:10.1016/S0266-3538(02)00105-7
2. Angel, G.D., Haritos, G.K.: An Immediate Formula for the Radius of Curvature of A Bimetallic Strip. International journal of engineering research and technology **2**, 1312–1319 (2013)
3. Both, J.C.: Tragfähigkeit von CFK-Metall-Laminaten unter mechanischer und thermischer Belastung: Zugl.: München, Techn. Univ., Diss., 2014. Ingenieurwissenschaften. Dr. Hut, München (2014). URL http://mediatum.ub.tum.de/node?id=1173300
4. Cowley, K.D., Beaumont, P.W.: The measurement and prediction of residual stresses in carbon-fibre/polymer composites. Composites Science and Technology **57**(11), 1445–1455 (1997). doi:10.1016/S0266-3538(97)00048-1
5. Crasto, A.S., Kim, R.Y.: On the Determination of Residual Stresses in Fiber-Reinforced Thermoset Composites. Journal of Reinforced Plastics and Composites **12**(5), 545–558 (1993). doi:10.1177/073168449301200505
6. Grunenfelder, L.K., Centea, T., Hubert, P., Nutt, S.R.: Effect of room-temperature out-time on tow impregnation in an out-of-autoclave prepreg. Composites Part A: Applied Science and Manufacturing **45**, 119–126 (2013). doi:10.1016/j.compositesa.2012.10.001
7. Hexcel Corporation: Product data sheet HexPly 8552 (2020). URL https://www.hexcel.com/user_area/content_media/raw/HexPly_8552_eu_DataSheet.pdf
8. Jeronimidis, G., Parkyn, A.T.: Residual Stresses in Carbon Fibre-Thermoplastic Matrix Laminates. Journal of Composite Materials **22**(5), 401–415 (1988). doi:10.1177/002199838802200502
9. Kappel, E., Stefaniak, D., Fernlund, G.: Predicting process-induced distortions in composite manufacturing – A pheno-numerical simulation strategy. Composite Structures **120**, 98–106 (2015). doi:10.1016/j.compstruct.2014.09.069
10. Khatkhate, A.M., Singh, M.P., Mirchandani, P.T.: A Parametric Approximation for the Radius of Curvature of a Bimetallic Strip. International Journal of Engineering Research and **V6**(06) (2017)
11. O'Brien, T.K.: Fatigue Delamination Behavior of PEEK Thermoplastic Composite Laminates. Journal of Reinforced Plastics and Composites **7**(4), 341–359 (1988). doi:10.1177/073168448800700403
12. Pagano, N.J., Hahn, H.T.: Evaluation of Composite Curing Stresses. In: G.M.L. Gladwell, J.N. Reddy (eds.) Mechanics of Composite Materials, *Solid mechanics and its applications*, vol. 34, pp. 57–69. Springer Netherlands, Dordrecht (1994). URL https://doi.org/10.1007/978-94-017-2233-9_6
13. Potter, K.D., Campbell, M., Langer, C., Wisnom, M.R.: The generation of geometrical deformations due to tool/part interaction in the manufacture of composite compo-

nents. Composites Part A: Applied Science and Manufacturing **36**(2), 301–308 (2005). doi:10.1016/j.compositesa.2004.06.002
14. Prussak, R., Stefaniak, D., Hühne, C., Sinapius, M.: Residual stresses in intrinsic UD-CFRP-steel-laminates - Experimental determination, identification of sources, effects and modification approaches. In: 20th Symposium on Composites, Materials Science Forum, pp. 369–376. Trans Tech Pubn (2015). doi:10.4028/www.scientific.net/MSF.825-826.369. URL `https://www.scientific.net/MSF.825-826.369.pdf`
15. Prussak, R., Stefaniak, D., Hühne, C., Sinapius, M.: Evaluation of residual stress development in FRP-metal hybrids using fiber Bragg grating sensors. Production Engineering **12**(2), 259–267 (2018). doi:10.1007/s11740-018-0793-4
16. Prussak, R., Stefaniak, D., Kappel, E., Hühne, C., Sinapius, M.: Smart cure cycles for fiber metal laminates using embedded fiber Bragg grating sensors. Composite Structures **213**, 252–260 (2019). doi:10.1016/j.compstruct.2019.01.079
17. Schulte, K.J., Hahn, H.T.: Prediction and Control of Processing-Induced Residual Stresses in Composites. Part 2: AS4/PEEK Composite, Pennsylvania State University Composites Manufacturing Technology Center: Final Technical Report. URL `https://apps.dtic.mil/sti/citations/ADA223696`
18. Schürmann, H.: Konstruieren mit Faser-Kunststoff-Verbunden. Springer Berlin Heidelberg, Berlin, Heidelberg (2007). doi:10.1007/978-3-540-72190-1
19. Stefaniak, D., Kappel, E., Spröwitz, T., Hühne, C.: Experimental identification of process parameters inducing warpage of autoclave-processed CFRP parts. Composites Part A: Applied Science and Manufacturing **43**(7), 1081–1091 (2012). doi:10.1016/j.compositesa.2012.02.013
20. Timoshenko, S.: Analysis of Bi-Metal Thermostats. Journal of the Optical Society of America **11**(3), 233 (1925). doi:10.1364/JOSA.11.000233
21. Unger, W.J., Hansen, J.S.: The Effect of Cooling Rate and Annealing on Residual Stress Development in Graphite Fibre Reinforced PEEK Laminates. Journal of Composite Materials **27**(2), 108–137 (1993). doi:10.1177/002199839302700201
22. Wiedemann, J., Mrzljak, S., Koord, J., Hühne, C., Walther, F.: Modified cure cycles for increased fatigue performance of fiber metal laminates. Composite Structures **351**, 118631 (2025). doi:10.1016/j.compstruct.2024.118631
23. Wiedemann, J., Schmidt, J.U.R., Hühne, C.: Applicability of asymmetric specimens for residual stress evaluation in fiber metal laminates. Journal of Composites Science **6**(11), 329 (2022). doi:10.3390/jcs6110329
24. Wisnom, M.R., Gigliotti, M., Ersoy, N., Campbell, M., Potter, K.D.: Mechanisms generating residual stresses and distortion during manufacture of polymer–matrix composite structures. Composites Part A: Applied Science and Manufacturing **37**(4), 522–529 (2006). doi:10.1016/j.compositesa.2005.05.019
25. Yu, Y., Ashcroft, I.A., Swallowe, G.: An experimental investigation of residual stresses in an epoxy–steel laminate. International Journal of Adhesion and Adhesives **26**(7), 511–519 (2006). doi:10.1016/j.ijadhadh.2005.07.006
26. Zobeiry, N., Poursartip, A.: The origins of residual stress and its evaluation in composite materials. In: P.W.R. Beaumont (ed.) Structural integrity and durability of advanced composites, Woodhead publishing series in composites science and engineering, pp. 43–72. Elsevier, Amsterdam (2015). doi:10.1016/B978-0-08-100137-0.00003-1

Chapter 6
Residual Stress Influence on Guided Ultrasonic Wave Propagation

Abstract For damage localization using GUW, the propagation velocity of the relevant wave is usually a necessary input variable. However, the propagation velocity of GUW is sensitive to applied loads or stresses. An experimental method is used to investigate how the manufacturing-induced residual stress state of an FML, which acts as a form of prestress in the material, affects the wave propagation velocity. This method allows the determination of dispersion diagrams over an extensive frequency range. The experimental results show that the method enables reproducible measurements across different specimen geometries. There is also good agreement between the dispersion relations for the fundamental A_0- and S_0-modes and analytical solutions. The method does not reveal any measurable influence of different residual stress states set by process modifications on the wave propagation velocity of the two fundamental modes in the frequency range investigated. Additional measurements under externally applied tensile prestress on the test specimens show that a prestress influence on the phase velocity also exists in FMLs. The known influence of a prestress on the wave propagation velocity demonstrates that the influence of the different residual stress states is in the same order of magnitude of the measurement reproducibility when using two different test specimens.[1 2]

6.1 Motivation

The sensitivity of GUW to damage is demonstrated in a preliminary study by introducing artificial delaminations during the manufacturing process of a specimen with layup ID2. The artificial delamination is realized by placing thin PTFE release

[1] Parts of the results of this chapter have been published in [4] and [22]

[2] The experimental method to determine dispersion diagrams using laser vibrometry and multifrequency excitation is based on the work of Tilmann Barth. The measurements in FML were performed in cooperation with Tilmann Barth and Natalie Rauter from the Institute of Mechanics of the Helmut Schmidt University/University of the Federal Armed Forces Hamburg and Thomas Roloff from the Institute of Mechanics and Adaptronics of the Technische Universität Braunschweig.

J. Wiedemann, *Relevance of Manufacturing-Induced Residual Stress States in Fiber Metal Laminates for Structural Health Monitoring*, Mechanics and Adaptronics, https://doi.org/10.1007/978-3-032-09925-9_6

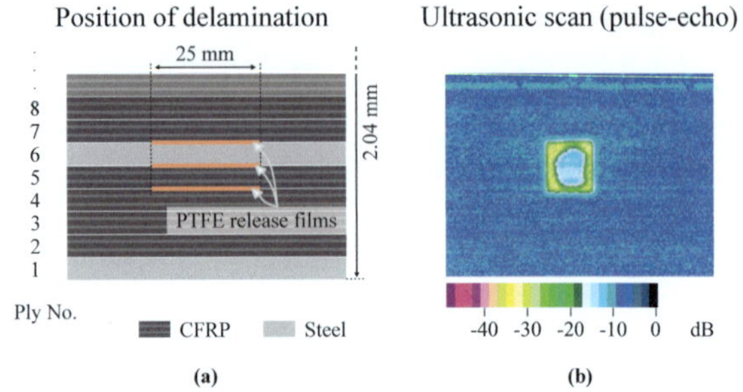

Fig. 6.1 **(a)** Square PTFE release films with an edge length of 25 mm are placed between plies 4-5, 5-6, and 6-7 of a specimen with layup ID2 to create artificial delaminations. **(b)** The ultrasonic scan (pulse-echo) proves the existence of a defect after manufacturing.

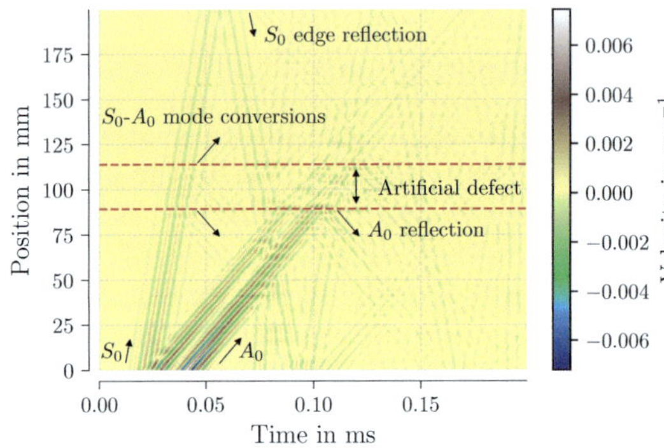

Fig. 6.2 Analyzing the B-scan image of the wave packets crossing the position of the artificial defect in the plate shows the interaction with fundamental GUW modes (reflection, mode conversion). The data is based on an LSV measurement with a five-cycle Hanning-windowed sine-burst excitation at 200 kHz.

films with a defined size (25 mm × 25 mm) at specified positions of the laminate. Figure 6.1 shows the positions of the release films in a cross-sectional view. An ultrasonic scan (pulse-echo mode)[3] of the manufactured plate proves that the release film patches can create a defect within the laminate.

The GUW wave interaction with the artificial defect in the specimen is monitored with a laser scanning vibrometer (LSV). GUWs are excited with a sine burst signal (five cycles, Hanning window) at a frequency of 200 kHz using a piezo-electric transducer attached to the specimen at a distance of 125 mm to the damage. The velocity on the plate surface is measured with the LSV in a straight line from the piezo across the defect. Figure 6.2 shows the corresponding B-scan image. The data indicates the interaction of the propagating wave packets with the damage. Mode conversion from the S_0-mode into the A_0-mode and significant reflections of the A_0-mode are visible.

In the B-scan image, the detected wave packets can be attributed to the two fundamental GUW modes (A_0 and S_0) based on their velocity. In practical SHM setups, such a large information density (B-scan) is usually unavailable. When using a pitch-catch setup with an actuator and a sensor, the time signals at only distributed locations are received. Such a signal is illustrated in Fig. 6.3 at the position 70 mm in

[3] The ultrasonic scans on specimens with artificial defects were performed in cooperation with Daniel Schmidt from the Institute of Lightweight Systems of the German Aerospace Center (DLR).

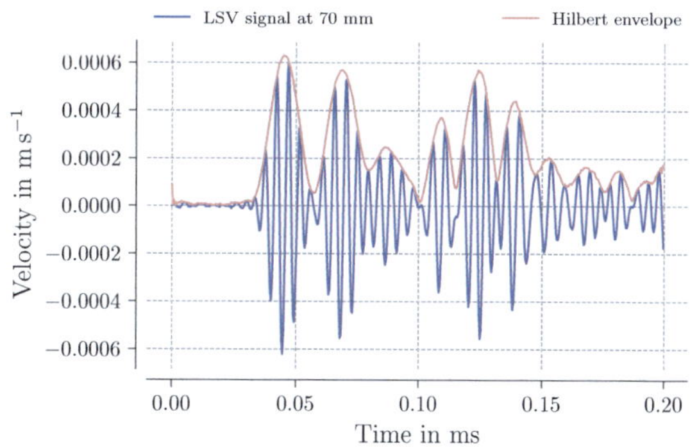

Fig. 6.3 LSV signal at position 70 mm in the measurement path (cf. Fig. 6.2) of the specimen with the artificial delamination.

the measurement path. From this information, it becomes challenging to differentiate between the single wave packets and to correlate the wave packets to reflections from damage, edges, or other structural features. Various methods exist to assign wave packets detected in such a setup with an actuator and one or more sensors to a location on the measured structure. Many of these methods rely on the propagation velocities of the relevant waves as input variables [19, 20] to locate the position of damage or a reflector in a more general case.

However, stresses within a material affect wave propagation velocities, a phenomenon known as the acoustoelastic effect [15, 16] as was already discussed in Chap. 2. Both theoretical considerations [11] and experimental studies [7] have confirmed this relationship for GUWs. The acoustoelastic effect has primarily been studied in homogeneous and isotropic materials, such as aluminum [3]. The research has shown that the influence of stress on the propagation velocity depends on the wave mode and frequency range. A particularly large influence of stress on the wave propagation velocity of fundamental GUW modes in isotropic materials is seen in the A_0-mode for low frequency-thickness products.

Therefore, in the context of SHM in FMLs, the question is whether the inherent residual stress state within an FML also affects the propagation velocity of GUWs in a significant way and whether this interaction affects damage localization methods. Therefore, this chapter will investigate the following hypothesis:

! Research hypothesis 3

Cure cycle-induced residual stress states influence the phase velocity of the fundamental GUW modes (A_0 and S_0) propagating in an FML.

Unlike the findings in the literature regarding isotropic materials, it is important to note that the residual stress state in FMLs differs significantly from a stress state in isotropic materials based on uniaxial prestress. The residual stress state in FMLs consists of simultaneous tensile and compressive stresses in different layers of the laminate (cf. Fig. 2.3).

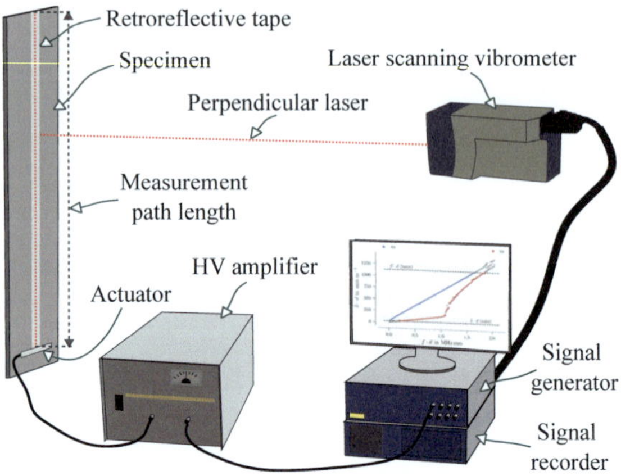

Fig. 6.4 For the measurement of GUW dispersion diagrams, an experimental setup with a laser scanning vibrometer is used (adapted from [22]).

6.2 Experimental Method to Determine Guided Ultrasonic Wave Dispersion Diagrams

Analyzing the effect of load on the GUW propagation velocity in the time domain is challenging [13]. Therefore, to analyze the GUW dispersion over an extensive frequency range, a method was originally proposed and used by [1,8,21], which also gave very good results in the work of Barth et al. [2] for isotropic materials. The method determines the structural velocity on the specimen's surface using an LSV. Time signals are recorded at points along a measuring path in the direction of the wave propagation to generate data as a function of time and spatial coordinates. The data are evaluated employing a non-uniform two-dimensional discrete Fourier transform (2d-DFT) to transform the measurement data from the time-spatial domain into the frequency-wavenumber domain [5]. The method is specially tuned to increase the accuracy of the measurements as much as possible by using multi-frequency excitation signals instead of impulse excitations while still enabling high automation and, thus, the possibility to cover large frequency ranges. This approach is based on the work of Barth et al. [2].

The experimental setup to measure the GUW propagation within a strip-shaped specimen is shown in Fig. 6.4. The GUWs are excited using a rectangular piezoelectric actuator (ceramic type: PIC 255, L $\times$ B $\times$ H: 30 mm $\times$ 5 mm $\times$ 0.2 mm) bonded to the specimen surface and an amplified multi-frequency excitation signal provided by a signal generator. With the LSV, the out-of-plane velocity after excitation is recorded on the specimen's surface, equipped with a retroreflective tape along the measurement path. Table 6.1 gives a detailed overview of the relevant hardware specifications of the devices used. To reduce edge reflections in the measured signal, all specimen edges are damped using an energy-dissipating butyl rubber tape.

The data evaluation in the frequency-wavenumber domain allows a more precise separation of the modes, of which at least two inevitably occur in the measurement

Table 6.1 Hardware specifications used in the experimental setup.

Device and parameter	Value
Laser scanning vibrometer	
Manufacturer	Polytec
Model	PSV-500
angular resolution	< 0.001°
Signal generator	
Manufacturer	Polytec
Bandwidth	30 MHz
No. of sampling points	1×10^6
Max. sampling rate	125 MHz
Used sampling rate	10 MHz
Signal recorder	
Manufacturer	Polytec
Sampling frequency	3.125 MHz
Used resolution	16 bit
Used amplitude range	100 mm/(sV)

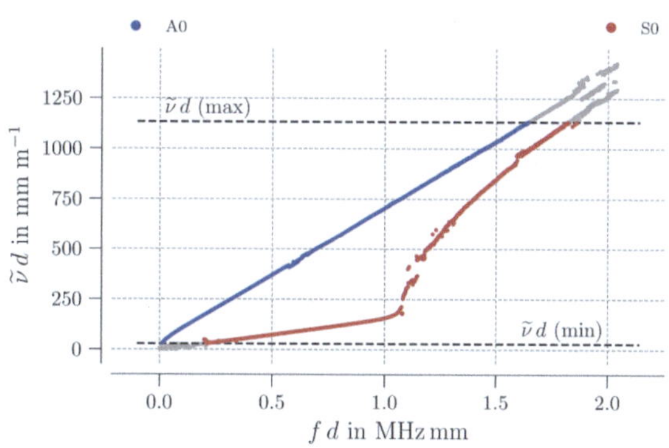

Fig. 6.5 The evaluable data range (colored data points) is limited by the minimum and maximum wavenumber, defined by the measurement path length and the distance between consecutive measurement points in the LSV measurement setup.

signal with the selected excitation. This is shown in Fig. 6.5 with the two fundamental modes A_0 and S_0, and well-separated data points.

Some limitations regarding the reliability of the data points exist. The measurement hardware and size of the specimen restrict the evaluable wavenumber range. A minimum of two measurement points per wavelength is necessary, restricting the evaluable data for large wave numbers by the spatial distance between two consecutive measurement points. For small wave numbers, the measurement path length is a restriction. Previous investigations [3] showed that a minimum of 20 wavelengths must fit inside the measurement path to ensure a reliable data evaluation, which restricts the measurement data towards small wave numbers. In this work, the data is analyzed down to a minimum of 10 wavelengths within the measurement path to ensure that effects that occur at small frequency-thickness products are not being overlooked. However, due to the expansion of this restriction, caution is required when evaluating the results for small frequency-thickness products.

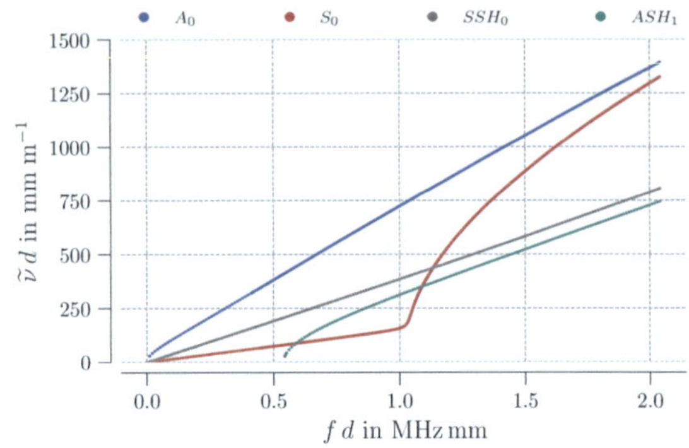

Fig. 6.6 The dispersion curves of the symmetric and antisymmetric shear horizontal GUW modes (SSH_0 and ASH_1) are crossing the S_0-mode path at around 1.1 MHz mm. Interference of the different GUW modes leads to difficulties in the data evaluation of the S_0-mode in this frequency-thickness range (adapted from [4]).

Additional limitations result from mode paths converging or crossing each other in the frequency-wavenumber representation. This can be seen at two locations in Fig. 6.5. For high frequencies, the wavenumbers of A_0- and S_0-mode converge, which results in a superposition of their amplitude maxima and, consequently, in an erroneous shift in the detection of both modes by the peak-search algorithm. In this case, the S_0-mode is affected more than the A_0-mode since the measured out-of-plane amplitudes for S_0 are considerably smaller than for A_0.

Another effect is the influence of the shear horizontal modes crossing the path of the S_0-mode in the dispersion diagram. This is illustrated in Fig. 6.6 based on analytical data. Despite the shear horizontal modes not having out-of-plane amplitudes, the interference with the S_0-mode can still be recognized in the signal.

All these effects lead to noise in the data and result in outliers being removed from the measured data. All data not considered for evaluation due to the discussed limitations is still plotted in grey (see Fig. 6.5) in all the subsequent figures as a reference.

In the following, all the dispersion diagrams are presented in the representation with phase velocity over frequency-thickness product, although the measurement data is evaluated in the frequency-wavenumber domain. The reason is the comparability to literature where the phase velocity representation is more common. The phase velocity c_p is defined as:

$$c_p = \frac{f}{\tilde{\nu}} \tag{6.1}$$

whereas f is the frequency and $\tilde{\nu}$ the corresponding wave number.

6.3 Validation of the Experimental Setup

This section compares different measurements to ensure that the experimental setup and evaluation approach produces reliable data when applied to FMLs. The first comparison involves an identical specimen measured twice using the same experi-

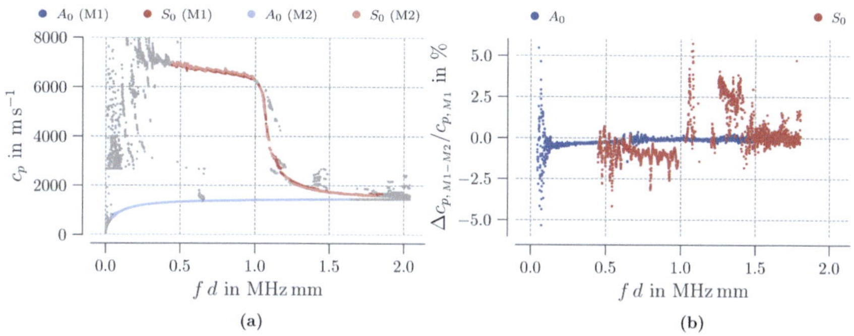

Fig. 6.7: **(a)** Comparison of the dispersion curves of A_0- and S_0-mode for two measurements (M1 and M2) with the identical experimental setup and specimen and **(b)** the relative phase velocity difference for each mode between the two measurements (adapted from [4]).

mental setup, with a new specimen installation between measurements. The second validation investigates the influence of the specimen geometry on the measurement results. A strip-shaped specimen and a plate-like specimen are compared. Finally, the measured dispersion diagrams are compared to analytically derived dispersion relations. All data in this section is based on the laminate layup ID2.

6.3.1 System Reproducibility

Figure 6.7a shows the dispersion diagram for two repeated measurements on an identical strip-shaped specimen (490 mm × 110 mm). The results show that the A_0-mode can be evaluated over a larger frequency range than the S_0-mode. The main reason for this is the smaller wavelength of the A_0-mode, which increases the evaluable frequency range for a given measurement path length. Moreover, the higher out-of-plane amplitude of the A_0-mode increases the signal-to-noise-ratio within the measurement data, contributing to fewer data points being detected as outliers. For small frequency-thickness ratios, the error within the A_0-mode significantly increases.

In Fig. 6.7b, the relative difference between the two measurements for each mode (A_0 and S_0) is plotted. The plot shows only slight deviations for both modes in the valid frequency range. Increased scattering, especially for lower frequencies, is due to increasing wavelength and the measurement path length. Higher deviations in the higher frequencies range result from reduced spatial resolution. The deviations in the S_0-mode at around 1-1.2 MHz mm result from the significant drop in velocity around this frequency-thickness product and the influence of the crossing shear horizontal waves. Nevertheless, the results display a good agreement between consecutive measurements on an identical specimen which confirms the robustness of the experimental measurement setup used for the analysis of the wave propaga-

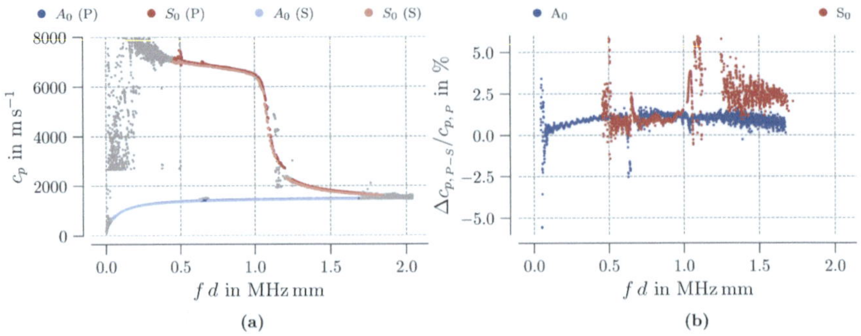

Fig. 6.8: (a) Comparison of the dispersion curves for the A_0- and S_0-mode for a plate-like specimen geometry (P) and a strip-shaped specimen (S) and **(b)** the relative phase velocity difference for each mode between the two specimen geometries (adapted from [4]).

tion in the frequency-wavenumber domain. The mean relative error between the two measurements over the entire frequency range is below 2% for both fundamental GUW modes (A_0 and S_0).

6.3.2 Specimen Geometry

In general, specimens with large in-plane dimensions are always favorable for investigations of wave propagation phenomena since wave packets reflected at the specimen edges are less relevant [17]. On the other hand, strip-shaped specimens with a considerably larger edge in the direction of the relevant wave propagation are beneficial for minimizing material use and manufacturing resources. Furthermore, the evaluations in Sec. 6.5 with specimens clamped in a tensile testing machine require strip-shaped specimens. Therefore, this section compares the dispersion diagrams for a strip-shaped specimen (490 mm × 110 mm) and a plate-like specimen (500 mm × 500 mm) with each other.

Figure 6.8 provides the corresponding dispersion curves for the two fundamental modes for both measurements and the difference in the phase velocities between the two specimen geometries. The differences between the two measurements are slightly higher than in the previous measurement on the identical specimen. Again the most considerable deviations occur for the S_0-mode in the critical frequency-thickness range (1-1.2 MHz mm).

Since the previous measurement was also performed on a strip-shaped specimen, the slight increase in deviations is attributed to using two different specimens instead of the specimen geometry. The mean difference over the entire frequency range for both fundamental modes is slightly higher than in the previous measurements on an identical specimen but still below 2% for both modes.

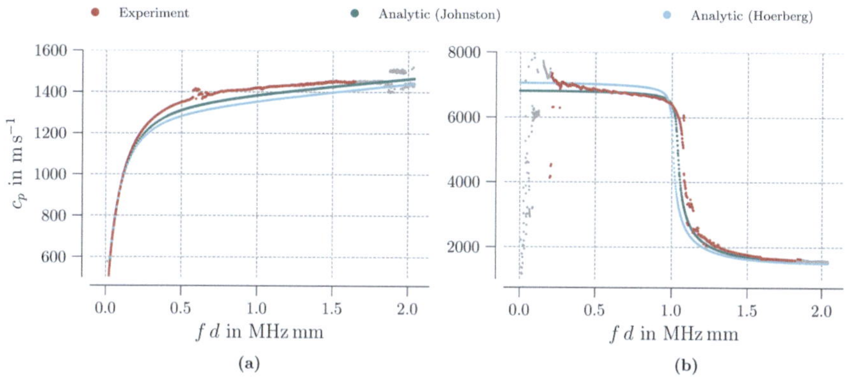

Fig. 6.9: Comparison of the measured dispersion relations with the analytic solutions shows good agreement for both modes: **(a)** A_0 and **(b)** S_0. Using two different material data sets for the CFRP (Johnston [12] and Hoerberg et al. [9]) shows the influence of material parameter uncertainties on the analytical solution. Note that the visible velocity ranges are limited to the relevant range for each mode (adapted from [4]).

Therefore, it can be concluded that strip-shaped specimens with an edge length of 110 mm (short edge) are still applicable for generating experimentally derived dispersion relations with the presented setup. Consequently, all subsequent experimental investigations use strip-shaped specimens with a width of 110 mm. It should be noted that the results cannot give a general recommendation of strip-shaped specimens and the usability is only proven for the frequency range investigated here. A further reduction of the specimen's width might lead to increased reflections and deviations in the measurement results.

6.3.3 Comparison to Analytical Solution

The software tool Dispersion Calculator (version 2.4) developed by [10] includes the possibility to define anisotropic and hybrid laminates. Therefore, the tool is used to compare the experimental data with results from the analytical solutions for guided waves in thin anisotropic laminates. The analytical solutions are calculated using the stiffness matrix method developed by [18]. For the analytical calculations, the material data presented in Chap. 3 are used. Since slight deviations between analytical and experimental data exist, two different material data sets are used in the analytic solution to illustrate the influence of the uncertainty in the material parameters. The two material data sets used are the ones from Johnston [12] and Hoerberg [9].

The comparison of the measurement data with the analytic solution shows very good agreement for the A_0-mode as shown in Fig. 6.9a. The results show that the material parameters used in the analytical calculation of the dispersion diagrams have a significant influence. The comparison of the two different material data sets reveals that using the CFRP parameters from [12] results in better agreement between analytics and measurement than for the parameters from [9]. The results are not intended to serve as a sensitivity study but should demonstrate the influence of material parameter uncertainties and the relevance of exact material parameters.

For the S_0-mode, the comparison between experimental and analytical values, as shown in Fig. 6.9b, reveals a similar pattern. Again, the solution derived with the material data set from [12] is closer to the experimentally derived dispersion curve, but again, minor deviations exist.

The agreement between the measured data and the analytically derived solutions for both fundamental modes proves that the measurement setup and experimentally determined dispersion curves are very accurate.

6.4 Comparison of Dispersion Diagrams for Different Residual Stress States

The investigations in this section are based on the validated experimental setup and the applicability of strip-shaped specimen geometries. To evaluate the influence of different residual stress states within otherwise identical specimens on the propagation velocities of fundamental GUW modes, the layups ID2 and ID4 are chosen. In layup ID2, high tensile residual stresses exist in the metal layers of the laminate. In contrast, layup ID4 possesses relatively high compressive stresses in the fiber layers due to the high metal volume fraction of the layup. By manufacturing two strip-shaped specimens of each layup with the two cure cycles MRCC and MOD, specimens with identical layups but high and low residual stress levels are realized.

Experimental and numerical results in isotropic materials show a significant influence of stresses on the wave propagation velocity of the A_0-mode at low frequency-to-thickness ratios [3]. Even if it is unclear whether these correlations also apply to residual stress states in FML, a particular focus is nevertheless placed on this frequency-thickness range in the evaluation. Therefore, the geometry of the test specimens is adapted compared to the previous measurements so that the influence of the residual stresses can be better resolved at small wavenumbers. The length of the strip-shaped specimens is increased to 900 mm compared to the previous investigations. The width of the specimens is kept constant at 110 mm.

The results of the measurements are shown in Fig. 6.10 for the specimens with layup ID2 and in Fig. 6.11 for the specimens with layup ID4. Both figures show the dispersion relations for the high (MRCC) and low (MOD) residual stress levels. From the phase velocity representations in the figures, no significant differences between the two stress states are visible. Therefore, the relative phase velocity differences between the two measurements are calculated and plotted on the right in the two figures. The differences between the two specimens with high and low residual stresses are of the same magnitude as those calculated between different specimen

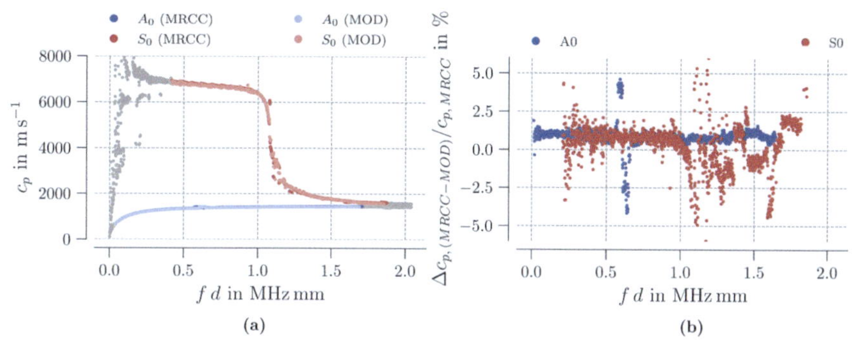

Fig. 6.10: **(a)** Comparison of the dispersion relations for the two different residual stress states (MRCC and MOD) of layup ID2 (MVF: 23.5%) and **(b)** corresponding relative phase velocity differences between the two specimens (adapted from [22]).

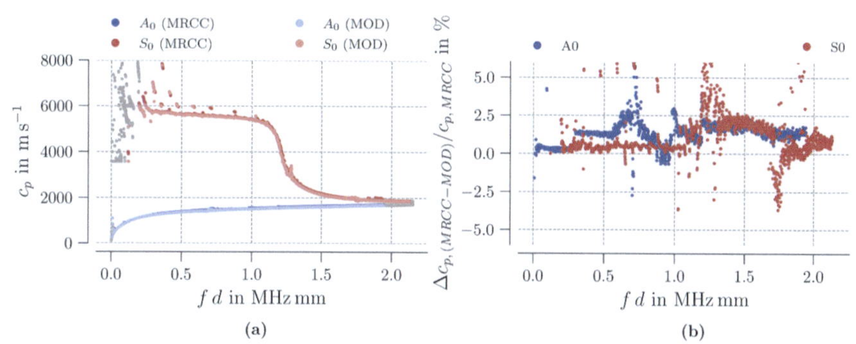

Fig. 6.11: **(a)** Comparison of the dispersion relations for the two different residual stress states (MRCC and MOD) of layup ID4 (MVF: 69.8%) and **(b)** corresponding relative phase velocity differences between the two specimens (adapted from [22]).

geometries and for repeated measurements on an identical specimen (cf. Figs. 6.7 and 6.8).

From the results, it can be concluded that no significant differences in the measured dispersion relations are observed, despite the substantial variation of the residual stress levels within the specimens with identical layups. The calculated phase velocity differences in the dispersion relations cannot be attributed to the difference in the residual stress state. The results indicate that the influence of the residual stress state on the GUW phase velocity is very small and cannot be measured with the given experimental setup. However, this conclusion is only valid for the two fundamental GUW modes investigated and for the frequency range measured.

6.5 Comparison of Dispersion Diagrams for Different Prestress States

The previous section showed that reducing the residual stress state by around 30% in the chosen specimens did not significantly affect the measured dispersion diagrams of the selected laminates. Therefore, an additional experiment is performed to further evaluate the influence of stress on wave propagation in FMLs. In this experiment, a strip-shaped FML specimen (layup ID2) is loaded with an uniaxial tensile load in the specimen's length direction using a tensile testing machine. The wave propagation in the specimen in its loaded state is measured with the same experimental setup illustrated in Fig. 6.4. Barth et al. [3] proved the method's feasibility for isotropic aluminum specimens under prestress.

In this work, the stress resulting from external loads is denoted as prestress to differentiate between the additional tensile stresses introduced by external loads and the residual stresses. It must be emphasized that with the materials in this work, it is impossible to manufacture residual stress-free laminates. Therefore, any prestresses represent additional stresses on top of an existing inherent residual stress state.

The tensile testing machine is force-controlled, and the force is calculated based on a nominal prestress. The nominal prestress is defined by the ratio of the tensile force and the nominal specimen cross-section. Due to the differences in the stiffness between steel and CFRP, nominal stress leads to different effective layer stresses in the steel and CFRP layers. This correlation is illustrated in Fig. 6.12. The nominal prestress is distributed across CFRP and steel layers based on their stiffness ratio. It must be superposed with the layerwise residual stresses to obtain the effective layer stresses in steel and CFRP.

In the following experiment, the nominal stresses are incrementally increased by steps of 50 MPa up to a maximum stress of 400 MPa, and the wave propagation is measured for each load step. Figure 6.12 shows the effective stress per layer in steel and CFRP for all load steps from 0 MPa to 400 MPa The maximum nominal stress level of 400 MPa results in an effective stress in the metal layer of 923 MPa. Compared to the ultimate tensile strength of 1121 MPa of the steel (cf. Table 3.1), the metal sheet stress is significant. However, compared to the nominal laminate strength of 2020 MPa (cf. Fig. 3.9) a nominal prestress of 400 MPa only accounts for around 20% of the laminate strength. This indicates that the purely linear elastic calculation of the effective layer stresses might no longer be accurate at high prestress levels, and stress transfer from the steel layers into the fiber layers might be relevant.

Figure 6.13 shows the dispersion relation for the A_0- and S_0-mode in the unloaded case and the case with the maximum prestress applied to the specimen (400 MPa). In the unloaded case, a minimal force of 150 N (0.7 MPa nominal prestress) was applied to the specimen in the tensile test machine to ensure a planar specimen orientation. This prestress state is denoted with 0 MPa in the following. In the figure, a difference in the measurements, particularly for the S_0 mode, is already seen.

To illustrate the effect of a prestress on the phase velocity, the relative phase velocity difference is calculated by:

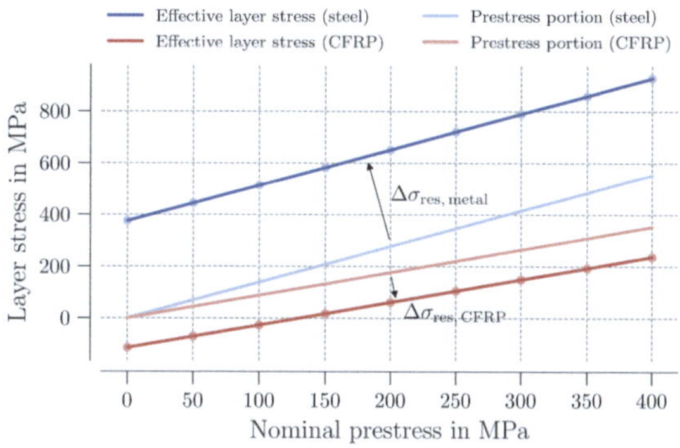

Fig. 6.12 The difference in stiffness between steel and CFRP leads to differences in the layer stress for a nominal prestress state and layup ID2. The effective layer stress results from the superposition of the prestress in each layer with its respective residual layer stress.

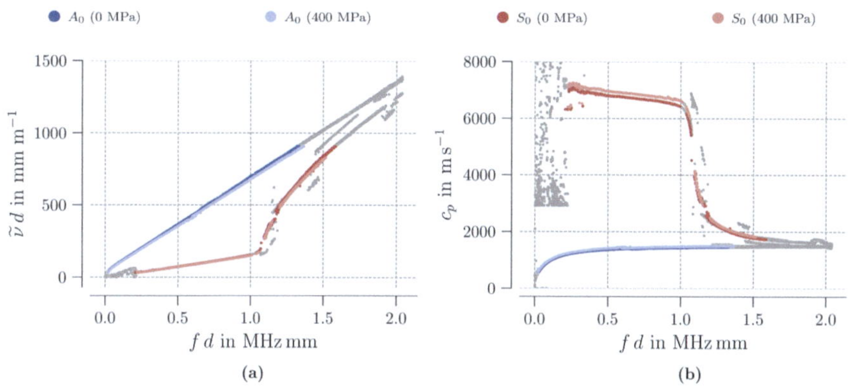

Fig. 6.13: Comparison of dispersion relations of A_0- and S_0-mode for minimum (0 MPa) and maximum (400 MPa) nominal prestress levels tested: **(a)** in wavenumber thickness representation and **(b)** phase velocity representation.

$$\Delta c_p = \frac{c_{p,\,\mathrm{x\,MPa}} - c_{p,\,0\,\mathrm{MPa}}}{c_{p,\,0\,\mathrm{MPa}}} \tag{6.2}$$

with x being the respective nominal stress. This is done for both fundamental modes A_0 and S_0 independently.

Figure 6.14 provides the relative difference in phase velocity of the A_0-mode for all stress levels tested. Comparing the relative phase velocity difference of all measurements, a considerable trend is observed, especially for small frequency-thickness products. This area is enlarged in Fig. 6.14b. The influence of a prestress on the phase velocity is qualitatively comparable to the results known in isotropic materials like aluminum, as was shown in [3].

Furthermore, the change of the phase velocity of the A_0-mode due to a prestress is linearly dependent on the nominal stress for a given frequency. This linear correlation is illustrated in Fig. 6.15 for three selected frequency-products. It can be seen that with increasing frequency, the correlation gets less clear, which can be mainly attributed

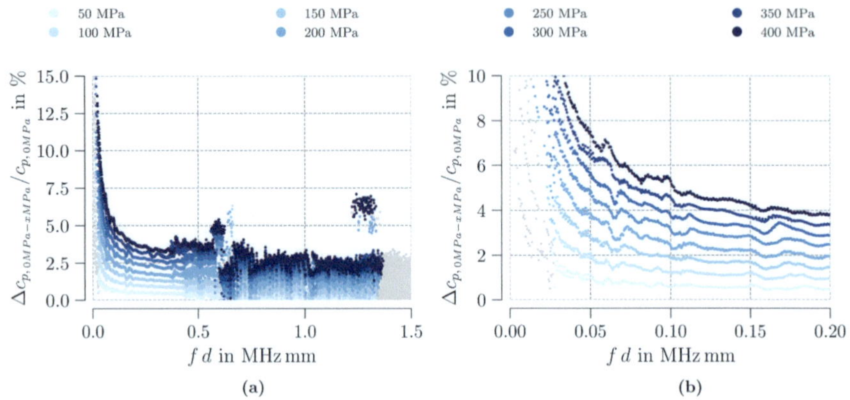

Fig. 6.14: Relative difference in phase velocity of the A_0-mode between the unloaded case and different nominal prestress levels: **(a)** evaluated frequency-thickness range and **(b)** zoom for small frequency-thickness products.

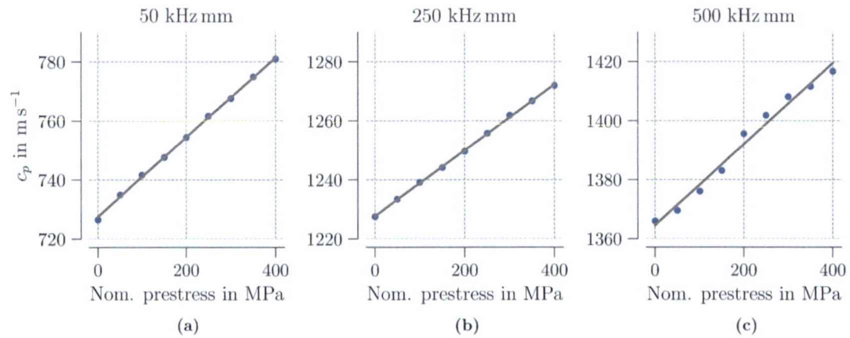

Fig. 6.15: The phase velocity of the A_0-mode increases linearly with a linear increase in nominal prestress: **(a)** at 50 kHz mm, **(b)** at 250 kHz mm, and **(c)** at 500 kHz mm. The data shows greater deviations from a linear approximation at higher frequency-thickness products.

to the decrease in the stress influence on the phase velocity. Again, this agrees with results found in aluminum [3].

Figure 6.16 provides the corresponding results for the S_0-mode. Again a linear correlation between prestress and the phase velocity difference can be observed for most parts of the evaluable frequency-thickness range. In contrast to the results in aluminum [3, 14], the evaluated frequency range shows an increase in phase velocity with increasing prestress. For aluminum, the non-linear elastic theory predicts a decrease in phase velocity with increasing prestress independent of the material model used (Neo-Hooke or Murnaghan) for small frequency-thickness products. This finding is interesting and needs further investigations in the future in combination with analytical or numerical models.

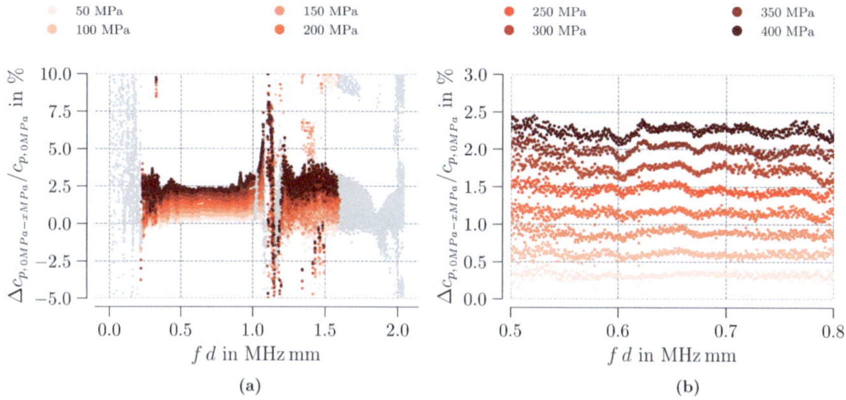

Fig. 6.16: Relative difference in phase velocity of the S_0-mode between the unloaded case and different nominal prestress levels: **(a)** evaluated frequency-thickness range and **(b)** zoom into the frequency-thickness range.

The following section discusses the results of the residual stress measurements in the context of the results found under prestress.

6.6 Conclusion and Consequences for SHM

The results in this chapter show that the method presented by Barth et al. [2] for experimentally determining dispersion diagrams over large frequency ranges using multi-frequency excitation signals and two-dimensional Fourier transformations of the measured data is also applicable to FMLs. The results further prove that the measurements can be repeated even across different specimens with high accuracy. The mean measurement error for repeated measurements on different FML specimens is below 2%.

The measured dispersion relations of the two fundamental GUW modes, compared with analytical data based on the stiffness matrix method and incorporated into the Dispersion Calculator software [10], show good agreement. Slight deviations exist in the range of uncertainties in the material property data from the literature.

The validated method investigated the influence of different residual stress levels in FML specimens. The GUW dispersion is measured in two CFRP-steel layups, one with relatively high residual tensile stresses in the metal sheets (375 MPa, MRCC, layup ID2) and one with relatively high compressive stresses in the fiber layers (−281 MPa, MRCC, layup ID4). With the modified cure cycle, the residual stress state in the specimens is reduced by around 30%. For the specimen with layup ID2, this results in a reduction of around 100 MPa in the residual tensile stresses in the

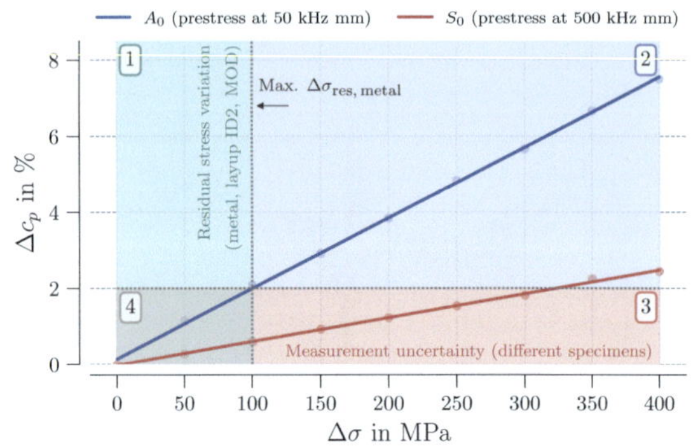

Fig. 6.17 The prestress results indicate that a change in residual stress of max. 100 MPa using MOD cycles and layup ID2 might affect the phase velocity by less than 2%, which is below the measurement uncertainty when using two different specimens.

metal sheets. The comparison of the experimentally derived dispersion relations for high and low residual stress states did not show any significant difference for the A_0- and S_0-mode in the investigated frequency range.

Additional measurements were carried out on FML laminates under prestress (tension) to place the results of the measurements under different residual stress levels in the context of the acoustoelastic effect. For this purpose, a specimen with layup ID2 was subjected to gradually increasing uniaxial tensile prestresses using a tensile testing machine. The results show an influence of prestress on the phase velocity for both the A_0- and S_0-modes. The influence is particularly noticeable at low frequency-thickness products in the A_0-mode and over the entire evaluable frequency-thickness range for the S_0-mode. The influence on the A_0-mode is qualitatively consistent with the results of the literature on isotropic materials. The phase velocity increases linearly with increasing prestress, but the effect gets smaller with increasing frequency-thickness products. For the S_0-mode, likewise, a positive linear relationship between the change in phase velocity and an increase in prestress is observed for most of the evaluated frequency-thickness range. The direction of this effect is opposite to the acoustoelastic effect as described by non-linear elastic theory in isotropic materials. The Neo-Hooke and Murnaghan material models describe a negative correlation between an increase in stress and a change in velocity for small frequency-thickness products in isotropic materials [3].

The most significant difference between measurements under varying residual stress states and under prestress is the need for two physical specimens to determine a residual stress influence. This is because the residual stress state can no longer be changed in a laminate after manufacturing (except by a change in temperature). In contrast, an identical laminate can be used to measure the influence of prestress on wave propagation. Therefore, the test setup and the specimen do not need to be changed to vary the amount of prestress, which reduces the measurement uncertainty significantly. As a result, the measurement under prestress is significantly less disturbed by material and manufacturing uncertainties, and smaller effect sizes can be resolved.

Fig. 6.17 compares the effect sizes and corresponding uncertainties to put the measurement results into context. This is done by showing the relative influence of a change in stress on the phase velocity. With the mean measurement uncertainty determined from measurements on two different physical specimens with identical layup and residual stress state and the maximum of the residual stress reduction in the metal sheets of layup ID2, the diagram in Fig. 6.17 can be divided into four segments. The mean measurement uncertainty on two different specimens is around 2%, while the maximum residual stress reduction in the metal sheets of layup ID2 using the MOD cycle is 100 MPa. Effect sizes found in segment (3) cannot be determined when two physical specimens are compared. The reason is the uncertainties due to the material, manufacturing process, and readjustment of the measurement setup, which accounts for around 2% change in phase velocity. Segment (1) indicates the range in which a residual stress state can be varied using the MOD cycle and the CFRP-steel specimen with layup ID2. A residual stress influence can be potentially measured if the effect is greater than 2% given the maximum change in residual stresses is 100 MPa. In segment (2), the influence of stress on the phase velocity can be measured, but the change in stress needs to be higher than achievable with MOD cycles. Finally, segment (4) illustrates the range in which a residual stress effect on the phase velocity cannot be measured due to the uncertainties when using two different specimens.

Adding the results of the prestress measurements for the A_0-mode at 50 kHz mm shows that a change in prestress of 100 MPa results in a phase velocity change of 2% at 50 kHz mm. This frequency-thickness product shows a particularly significant influence of a prestress on the A_0-mode and is at the lower end of the evaluable measurement range in this setup. For the S_0-mode, the influence is shown at a frequency-thickness product of 500 kHz mm. The results of the S_0 mode are less dependent on frequency, and the same effect size is seen over most of the evaluable frequency-thickness range. With the prestress results as a reference, it is shown that the expected change in phase velocity due to a change in (residual) stress of 100 MPa is below the measurement uncertainty for both modes.

It must be noted that the mean measurement error when using two specimens is calculated over the entire evaluable frequency range. The error even increases for lower frequencies due to the small wave numbers of the A_0-mode for a given specimen length. Furthermore, in contrast to a tensile prestress applied by a tensile testing machine, the residual stress state consists of both tensile (metal) and compressive (FRP) stresses. For simplicity, since the tensile residual stresses in the metal sheets are significantly higher than the compressive stresses in the fiber layers for layup ID2, only this portion of the residual stresses is compared in Fig. 6.17. It is unclear whether the combination of tensile and compressive residual stresses within FMLs has opposing effects resulting in an even less significant influence than assumed in the figure.

To investigate the effects further, future work should include analytical or numerical models and extended measurements on other FML specimen configurations to provide a fundamental understanding of the influence of prestress and residual stress on GUW propagation. To increase the accuracy in measurements of a residual stress influence, the use of a single specimen at varying temperature levels could solve this challenge. By varying the temperature, significantly higher residual stress differences can be realized within a single specimen. However, since GUW propagation is also sensitive to changes in temperature [6], it will be challenging to distinguish between a temperature effect and a stress effect.

The presented results allow assessing the third research hypothesis of this work, which states that *cure cycle-induced residual stress states influence the phase velocity of the fundamental GUW modes (A_0 and S_0) propagating in an FML.* Changing the residual stress state by modifying the curing process does not lead to any measurable change in the wave propagation velocity of the two fundamental modes A_0 and S_0 in the frequency range investigated. The influence seems smaller than the uncertainty introduced by the material, specimen, and manufacturing setup. Therefore, the hypothesis is rejected with the limitation of the residual stress difference that can be achieved by modified cycles regardless of a temperature influence on the residual stress state.

The results can now be related to the accuracy of damage localization using an SHM system. Assuming that the influence of the residual stress state on the wave propagation velocity is less than or equal to 2%, and assuming a linear relationship between propagation speed, propagation length, and propagation duration, this leads to an error of 2% or less in the localization accuracy. Simplified, this means that for a propagation path length of 1000 mm between the actuator, reflector, and sensor, the deviation in determining the position of the damage is below 20 mm. However, by using several sensors, the error can be reduced to a certain extent if the intersection is formed from several sensor signals from different directions. It is, therefore, concluded that the influence of residual stress due to cure cycle modifications in damage localization with the aid of the two fundamental modes is of subordinate importance. However, the results under prestress show a significant influence of an externally applied stress state with a change in phase velocity up to 10% in the frequency range often used for damage detection and localization. Therefore, it can be concluded that the influence of stress in general cannot be ignored in SHM systems used for FML structures.

References

1. Alleyne, D., Cawley, P.: A two-dimensional Fourier transform method for the measurement of propagating multimode signals. The Journal of the Acoustical Society of America **89**(3), 1159–1168 (1991). doi:10.1121/1.400530
2. Barth, T., Rauter, N., Lammering, R.: Experimental Determination of Lamb Wave Dispersion Diagrams Using 2D Fourier Transform and Laser Vibrometry. Preprint available at Research Square (2022). doi:10.21203/rs.3.rs-1321459/v1
3. Barth, T., Rauter, N., Lammering, R.: Experimental and numerical investigations on the acoustoelastic effect in hyperelastic waveguides. Structural Health Monitoring (2024). doi:10.1177/14759217241246506
4. Barth, T., Wiedemann, J., Roloff, T., Behrens, T., Rauter, N., Hühne, C., Sinapius, M., Lammering, R.: Experimental determination of dispersion diagrams over large frequency ranges

for guided ultrasonic waves in fiber metal laminates. Smart Materials and Structures **32**(8), 085011 (2023). doi:10.1088/1361-665X/ace0ea

5. Bracewell, R.N.: The Fourier transform and its applications, second edition, revised. edn. McGraw-Hill series in electrical engineering. McGraw-Hill, New York and St. Louis and Lisbon and London (1986)
6. Dodson, J.C., Inman, D.J.: Thermal sensitivity of Lamb waves for structural health monitoring applications. Ultrasonics **53**(3), 677–685 (2013). doi:10.1016/j.ultras.2012.10.007
7. Gandhi, N., Michaels, J.E., Lee, S.J.: Acoustoelastic Lamb wave propagation in biaxially stressed plates. The Journal of the Acoustical Society of America **132**(3), 1284 (2012). doi:10.1121/1.4740491
8. Hora, P., Červená, O.: Determination of Lamb wave dispersion curves by means of Fourier transform. Appl. Comput. Mech. **2012**(6), 5–16 (2012)
9. Hörberg, E., Nyman, T., Åkermo, M., Hallström, S.: Thickness effect on spring-in of prepreg composite L-profiles – An experimental study. Composite Structures **209**(8), 499–507 (2019). doi:10.1016/j.compstruct.2018.10.090
10. Huber, A.: Numerical Modeling of Guided Waves in Anisotropic Composites with Application to Air-coupled Ultrasonic Inspection. Dissertation, Universität Augsburg (2021)
11. Husson, D.: A perturbation theory for the acoustoelastic effect of surface waves. Journal of Applied Physics **57**(5), 1562–1568 (1985). doi:10.1063/1.334471
12. Johnston, A.A.: An integrated model of the development of process-induced deformation in autoclave processing of composite structures. Dissertation, University of British Columbia, Vancouver, Kanada (1997). doi:10.14288/1.0088805
13. de Luca, A., Minardo, A., Aversano, A., Vallifuoco, R., Sepe, R., Caputo, F.: Health and load monitoring in an aluminium plate through guided waves. Procedia Structural Integrity **52**, 424–429 (2024). doi:10.1016/j.prostr.2023.12.043
14. Mohabuth, M., Kotousov, A., Ng, C.T.: Effect of uniaxial stress on the propagation of higher-order Lamb wave modes. International Journal of Non-Linear Mechanics **86**(9), 104–111 (2016). doi:10.1016/j.ijnonlinmec.2016.08.006
15. Pao, Y.H., Gamer, U.: Acoustoelastic waves in orthotropic media. The Journal of the Acoustical Society of America **77**(3), 806–812 (1985). doi:10.1121/1.392384
16. Qu, J., Liu, G.: Effects of Residual Stress on Guided Waves in Layered Media. In: D.O. Thompson, D.E. Chimenti (eds.) Review of Progress in Quantitative Nondestructive Evaluation, vol. 45, pp. 1635–1642. Springer US, Boston, MA (1998). doi:10.1007/978-1-4615-5339-7_212
17. Rittmeier, L., Roloff, T., Rauter, N., Mikhaylenko, A., Haus, J.N., Lammering, R., Dietzel, A., Sinapius, M.: Influence of a Flat Polyimide Inlay on the Propagation of Guided Ultrasonic Waves in a Narrow GFRP-Specimen. Materials (Basel, Switzerland) **15**(19) (2022). doi:10.3390/ma15196752
18. Rokhlin, S.I., Wang, L.: Stable recursive algorithm for elastic wave propagation in layered anisotropic media: stiffness matrix method. The Journal of the Acoustical Society of America **112**(3 Pt 1), 822–834 (2002). doi:10.1121/1.1497365
19. Shan, S., Qiu, J., Zhang, C., Ji, H., Cheng, L.: Multi-damage localization on large complex structures through an extended delay-and-sum based method. Structural Health Monitoring **15**(1), 50–64 (2016). doi:10.1177/1475921715623358
20. Sharif-Khodaei, Z., Aliabadi, M.H.: Assessment of delay-and-sum algorithms for damage detection in aluminium and composite plates. Smart Materials and Structures **23**(7), 075007 (2014). doi:10.1088/0964-1726/23/7/075007
21. Su, Z., Ye, L.: Identification of damage using Lamb waves: From fundamentals to applications, *Lecture notes in applied and computational mechanics*, vol. 48. Springer, Berlin (2009)
22. Wiedemann, J., Barth, T., Roloff, T., Kluge, T., Rauter, N., Hühne, C.: Influence of residual stresses on the dispersion behavior of guided ultrasonic waves in fiber metal laminates. In: Proceedings of the 14th International Workshop on Structural Health Monitoring. Destech Publications, Inc (2023). doi:10.12783/shm2023/36729

Chapter 7
Residual Stress Influence on Fatigue Life

Abstract To investigate the influence of cure cycle-induced residual stresses on the fatigue life in CFRP-steel laminates, experimental fatigue tests with a constant amplitude and tension-tension loading are performed. The notched FML specimens are monitored with digital image correlation, thermography, and electrical resistance measurements throughout the dynamic tests. The evaluation of the test data shows that cracks are initiated at the stress concentrations around the edge of the notch and propagate through the metallic facesheets. Thermography reveals that failure of the inner metal sheets happens considerably after the failure of the facesheets. The influence of different residual stress states in otherwise identical specimens is seen in the evaluation of the dynamic stiffness and, in particular, when evaluating the number of cycles until crack initiation and facesheet failure. With reduced residual stresses, crack initiation is delayed, and crack growth is slowed down. The importance of the residual stresses for the fatigue life characterization becomes evident when the prediction bounds for the fit of the experimental crack initiation and crack propagation data are calculated. These prediction bounds become considerably smaller when the correct residual stress states for each laminate are considered in the calculation. Fatigue in the investigated CFRP-steel laminates is governed by the metal, and the propagation of a fatigue crack takes up most of the time of the fatigue life. This is particularly advantageous for an SHM system in FMLs.[1][2]

7.1 Motivation

The Federal Aviation Administration (FAA) [6] advises aircraft's principal structural components to be designed in a damage-tolerant manner with slow crack growth rates and the possibility to detect damage before the residual strength of the component

[1] Parts of the results of this chapter have been published in [31] and [32]

[2] The experiments were performed in cooperation with Selim Mrzljak from the Chair of Materials Test Engineering (WPT) of the TU Dortmund University.

J. Wiedemann, *Relevance of Manufacturing-Induced Residual Stress States in Fiber Metal Laminates for Structural Health Monitoring*, Mechanics and Adaptronics, https://doi.org/10.1007/978-3-032-09925-9_7

falls below a critical threshold. The extended fatigue life of FMLs after macro-crack initiation compared to monolithic metals proves their damage tolerance and it allows for larger inspection intervals and increased flexibility for condition-based maintenance (CBM) [1, 7]. To implement CBM effectively in an SHM system, a thorough understanding of the material's fatigue behavior is essential for predicting the remaining service life after detecting damage.

Fatigue in FML is governed by the metal constituent since metals are generally more susceptible to damage from cyclic fatigue loading than composites [1]. In fatigue life analysis, it must be differentiated between initiation-controlled damage development and propagation-controlled damage development [23]. Micro-crack initiation governs fatigue in monolithic metals because macro-crack growth is fast and only relevant to a small proportion of fatigue life [9]. Although the metallic component initiates the failure in FML, the relevant phenomenon observed is not crack initiation as in monolithic metals but macro-crack propagation. This is because, in FMLs, a constant crack growth rate prevails over an extensive range of the fatigue life, which can be explained by the number of interfaces between metal and FRP and the fiber bridging mechanism [1,25]. The surrounding fiber layers reduce stress peaks at the crack tip in the metal, as the fibers take over the load-bearing function with increasing crack size. Moreover, the layered structure and the increasing elongation of the fiber layers cause the crack plane to deflect into the laminate plane and progress as delaminations between the fiber and metal layers. Overall, the fatigue process is determined by a continuous mutual interaction between crack length, crack opening, and delamination. This leads to the superior fatigue damage tolerance of FMLs compared to monolithic metals and is a primary reason for the use of GLARE in the upper fuselage of the Airbus A380 passenger aircraft [3, 30].

For GUW-based SHM systems, this fatigue behavior of FMLs is advantageous. Micro-cracks with geometric dimensions below the wavelength of GUWs in the typical frequency range used for damage detection are expected to have a minor influence on the evaluated features due to their low interaction with propagating waves. Therefore, they are challenging to detect. A possible solution is the evaluation of the non-linearity of the GUW propagation that is sensitive to the accumulation of micro-damage in FRP, as shown in [22]. However, in FRPs, microcrack accumulation is a distributed phenomenon, whereas, in metals, fatigue damage is initiated at specific locations with high stress concentrations. When focusing on macro-crack detection and propagation, the crack length is of the order of the wavelength of the GUWs. Hence, a more considerable interaction between damage and propagating waves is expected.

The state of the research in Chap. 2 shows that residual stresses contribute significantly to the fatigue life of an FML component. However, this was experimentally shown mainly for GLARE laminates using the post-stretching method [10]. The difference in the residual stress states achieved with modified cure cycles is smaller than with post-stretching, but a difference in fatigue behavior is still seen [12]. The literature shows some inconsistencies in the experimental results regarding the influence of the cure cycle on fatigue life. Furthermore, most of the investigations are on GLARE [15], CARALL [11] or FRP [12], while [8] is one of the few to use a combination of CFRP and steel. Therefore, a comprehensive investigation of the cure cycle influence on fatigue in CFRP-steel laminates is justified to test for the following hypothesis:

! Research hypothesis 4

The reduction of cure cycle-induced residual stresses improves the fatigue life of CFRP-steel laminates under constant amplitude tension-tension fatigue loading.

7.2 Experimental Setup and Evaluation

Fatigue cracks in metals originate from inhomogeneities in the material. Hence, to ensure that the fatigue cracks initiate within the observed region of a specimen, open hole tension (OHT) specimens with the hole serving as a starter notch are manufactured. Furthermore, the chosen laminate configurations provided in Table 3.5 with metal sheets as outer layers allow the initiation and propagation of fatigue cracks to be observed during experimental testing in the metal. The experiments focus on constant amplitude tension-tension fatigue, a relevant loading condition, for example, in the pressurization cycle of an aircraft fuselage.

The rectangular fatigue test specimens with a size of 180 mm x 20 mm and the 5 mm drill hole are illustrated in Fig. 7.1. The geometry of the fatigue specimens has a width-to-diameter (w/d) ratio of 4, which proved its usability in other investigations [19, 29] and was chosen based on realistic applications (e.g., rivet hole) and to keep material consumption in a reasonable range. Moreover, a w/d-ratio of 4 is still covered by analytical solutions for quasi-static OHT of composite materials [13], which allows the correlation of the locations for crack initiation with analytical solutions. Specimens with all four different layup configurations (cf. Table 3.5) are manufactured and tested. Using these configurations, a laminate thickness range of 2.04 mm to 6.52 mm, a range of the MVF of 20.2% to 69.8%, a range in the NoI of 6 to 20 and a range in the FTF of 3.7% to 23.3% is achieved. This large laminate variation ultimately allows the influence of the layup and various residual stress levels on the fatigue behavior to be evaluated.

The fatigue tests are performed with a Shimadzu EHF-EV50 (Shimadzu Corporation, Kyoto, Japan) servo-hydraulic testing system with a load capacity of 50 kN. During the tests, a three-dimensional digital image correlation (DIC) system (Limess Q400 3D-DIC, Limess Messtechnik und Software GmbH, Krefeld, Germany) with 5 MP cameras is triggered at maximum stress to determine the crack length over time. Four cameras are used: two in the front and two in the rear (Ricoh TV lens FL-CC7528-2M, 75 mm focal length).

The temperature of the specimens is monitored with an Infratec ImageIR 8800 (InfraTec GmbH Infrarotsensorik und Messtechnik, Dresden, Germany) thermographic camera. A high-precision Keithley (Tektronix Inc., Beaverton, USA) system, con-

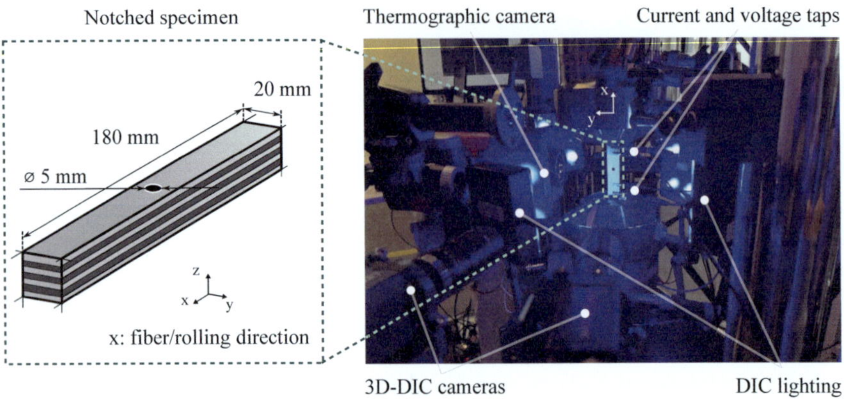

Fig. 7.1: Specimen geometry and photo of the test setup during fatigue testing. The specimen thickness is exaggerated in the specimen sketch and varies depending on the laminate configuration (Table 3.5) between 2.04 mm and 6.52 mm (adapted from [32]).

taining a 2601B source measure unit connected via R232 null modem cable to a 2182A nanovoltmeter (Tektronix Inc., Beaverton, USA), records the electrical resistance across the metal facesheets. With increasing crack size, the electrical resistance increases significantly and eventually goes to infinity when the facesheet fails. With the electrical resistance measurements, crack initiation and propagation are detected. The system is operated in delta mode to compensate for thermo-electrical effects [19]. To achieve reproducible measurement results, markings with a distance of 50 mm are painted onto the specimens to place the voltage drop clamps and the current in/out clamps at identical locations on all specimens. Figure 7.1 shows a photo of the fatigue testing setup.

Constant amplitude tests (CATs) are conducted under ambient temperature with a sinusoidal load-time function at a stress ratio R = 0.1 (tension-tension loading). The laminates are tested under two different maximum stress levels (400 MPa and 600 MPa). One exception is the ID3a laminates, which are only tested for a maximum stress level of 400 MPa due to the load limit of the 50 kN testing machine.

The maximum stress levels are chosen based on the stress-strain data of the steel material (Table 3.1). A representative stress-strain curve for the steel sheet with a thickness of 0.12 mm is shown in green in Fig. 7.2. Linear elastic interpolation of the initial slope shows that the first plastification processes start from approx. 600 MPa. A load below 64% of the elastic limit did not lead to any damage in fatigue tests up to 1×10^6 cycles of CFRP-steel laminates in [4]. In the regime of very high cycle fatigue ($\geq 1 \times 10^7$ cycles), cracks can occur even at loads lower than the elastic limit, where special high-frequency fatigue testing systems are used for timely efficient characterization [21]. However, concerning possible applications of the CFRP-steel laminate in this work, the focus is on the low and high cycle fatigue regime ($\leq 1 \times 10^6$ cycles). Therefore, even though all the laminates have different stiffnesses due to their layup architecture, the two upper-stress levels of 400 MPa and

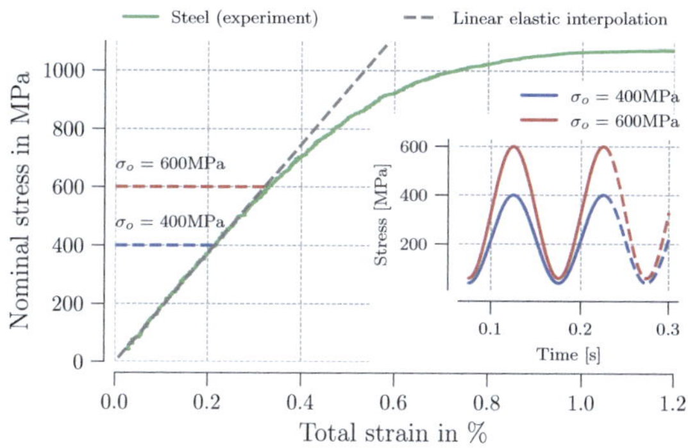

Fig. 7.2 Definition of the two upper stress levels σ_o for fatigue testing based on the quasi-static stress-strain data of the steel material and corresponding stress cycles during continuous amplitude testing (adapted from [32]).

600 MPa are selected as a rough estimate. The stresses refer to the notched cross-section of the test specimens during the CATs. The testing frequency is set to 10 Hz for all experiments to avoid excessive testing time but also limit self-heating [18].

The evaluations of the DIC and electrical resistance measurements are used to determine the cycle numbers of crack initiation and facesheet tear in the specimens. Crack initiation is defined by a significant change in electrical resistance combined with a visible crack in the DIC images. The thermographic images are used to assess the state of the inner metal sheets. Furthermore, the force (F) and displacement (s) data of the testing machine, measured by load cell and linear variable differential transformer, is used to evaluate the dynamic specimen stiffness over its fatigue life. The dynamic stiffness, which is derived from the force-displacement hysteresis, represents the damage accumulation concerning the load cycle. Based on the force-displacement data, the dynamic stiffness C_{dyn} is formulated instead of E_{dyn} as in [20, 29]:

$$C_{\text{dyn}} = \frac{F_{\text{max}} - F_{\text{min}}}{s_{\text{max}} - s_{\text{min}}} \tag{7.1}$$

The experimental results are mainly discussed with the fatigue results obtained during the tests with a maximum stress level of 400 MPa, while similar conclusions can be drawn when comparing the results of the tests with the higher stress level of 600 MPa.

7.3 Experimental Fatigue Damage Evolution in CFRP-Steel Laminates

This section illustrates the evolution of fatigue damage observed for the tested FML configurations, using the specimen with layup ID6b manufactured with the standard cure cycle (MRCC) as a representative example.

Figure 7.3 shows the two-dimensional strain fields in loading direction (ε_{xx}) of specimen ID6b at a stress level of 400 MPa. The experimentally determined strain

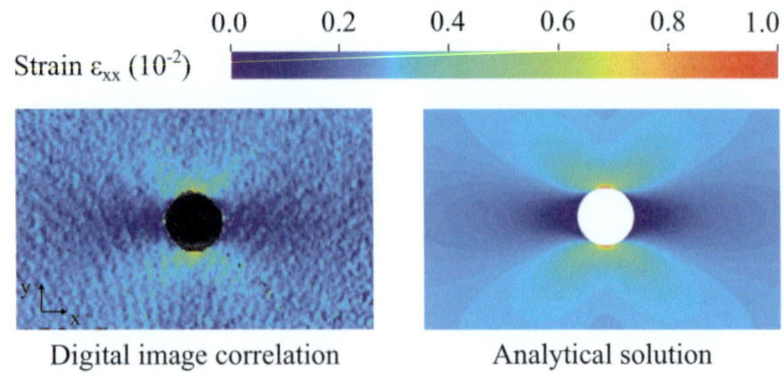

Fig. 7.3 Surface strains from digital image correlation measurements agree well with analytical solutions at maximum stress during fatigue loading for layup ID6b and $\sigma_{\max}$ = 400 MPa (adapted from [32]).

field through DIC measurement is compared to the analytically determined strain field around the center notch. It should be noted that the DIC results show the surface strain of the specimen, while the analytical result represents the midplane strain. However, the results are comparable due to the symmetry of the laminate and the fact that the specimen is not loaded in bending. The analytical solution is based on the Lekhnitskii formalism [16] for an orthotropic elastic plate with a circular center notch in loaded tension. The derivation of the Lekhniskii formalism is frequently covered in literature [14, 24, 26] and its accuracy has been demonstrated [13]. More details can be found in [32].[3]

The two strain fields show a remarkable level of agreement. The pattern of the strain as it decreases from the concentration at the center towards the outer regions of the segment is almost identical in both strain fields. Similarly, the peaks around the center hole are also very similar. This high level of agreement confirms the accuracy and validity of the DIC measurement, which is essential for further analysis.

Based on the strain field characteristics, the most probable location for initiating fatigue cracks is the strain concentration at the notch's edge. Although the strain field analysis assumes the cracking initiation point, the crack growth behavior must be experimentally determined.

All fatigue test specimens are tested up to a cycle number of 6×10^5, except for the specimens with ID4 that failed at around 1×10^5 ($\sigma_{\max}$ = 400 MPa) and 3×10^4 ($\sigma_{\max}$ = 600 MPa) cycles respectively. The reason is the high FTF in layups ID4, which leads to laminate failure after facesheet tear since the remaining layers lack enough residual strength. No final laminate failure occurred within 6×10^5 cycles for all other layups.

A similar failure damage pattern was found for all specimens independent of the layup and cure cycle. Images of the DIC cameras reveal the progression of the damage over the number of cycles during fatigue testing for the steel facesheets. Figure 7.4 shows the strains (DIC) in the loading direction (ε_{xx}) on the ID6b specimen front surface for specific load cycles. As indicated by Fig. 7.3 a crack is initiated in the facesheets at the notch's edge. The cracks propagate through the facesheets and transverse to the loading direction until the complete metal sheet

[3] The analytical solutions are based on the work of Josef Koord from the Institute of Lightweight Systems of the German Aerospace Center (DLR).

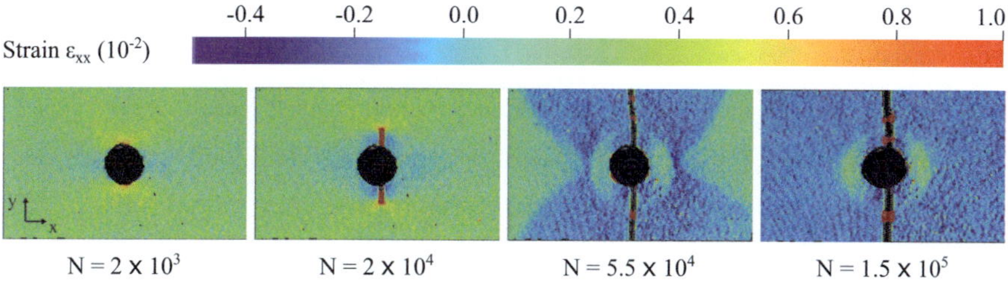

Fig. 7.4: The surface strains in the top metal layer measured with digital image correlation of specimen ID6b MRCC (σ_{max} = 400 MPa) show the evolution of the damage in the metal facesheet (crack initiation, crack propagation with simultaneous delamination and subsequent facesheet failure). Note that the scale of the color legend differs from Fig. 7.3 to increase the visibility of the delaminated areas (adapted from [32]).

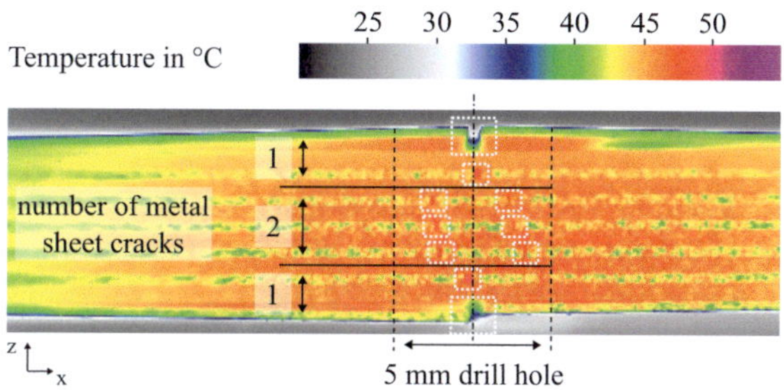

Fig. 7.5 Thermographic image of the side view of specimen ID6b MRCC (σ_{max} = 400 MPa) reveals a significant temperature increase and inner metal sheet failure after 5.5×10^5 cycles (adapted from [32]).

tears. Crack growth is accompanied by delamination of the facesheets. Simultaneous damage accumulation is observed for the front and rear of the specimen.

The DIC images can only show damage in the facesheets. However, the thermographic cameras focus on the side of the test specimens and reveal that the inner metal sheets only tear after facesheet failure (Fig. 7.5). A phenomenon also observed by Burgers and Kempen [5] that showed a significant difference in crack length between facesheet cracks and cracks in the inner metal sheets can occur. One possible reason could be that the facesheets are only supported by bridging fibers from one side in contrast to the metal sheets inside the laminate [2]. In contrast, other investigations by Takamatsu et al. [28], show that the crack lengths between the surface aluminum sheets and the center aluminum sheets in GLARE can also be of similar size.

While the two inner steel sheets closest to the facesheets also fail with a single crack each, two cracks per sheet occur in the three center steel sheets of specimens with ID6b. The dark red spots in Fig. 7.5 indicate the cracks, which are locations with higher temperatures due to the increased friction of the crack faces under cyclic load. Multiple cracks in the metal sheets of FML under cyclic loading have also been observed in [19], where the number of cracks increased with increasing load level. Here, it can be assumed that after facesheet failure, the stress concentrations at the notch of the inner metal sheets will increase, leading to a similar phenomenon.

The evaluation of the dynamic stiffness, defined in Eq. (7.1), provides further insights into damage progression in addition to looking at the imaging techniques.

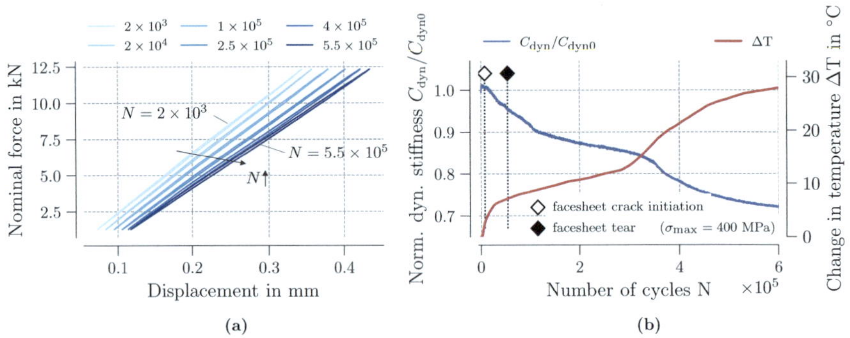

Fig. 7.6: (a) Load hystereses of specimen ID2 (MRCC, σ_{max} = 400 MPa) for specific numbers of increasing cycles. **(b)** Normalized dynamic stiffness and change in maximum specimen temperature over the number of cycles of specimen ID6b (MRCC, σ_{max} = 400 MPa) during cyclic loading (adapted from [32]).

Figure 7.6a shows the force-displacement data for selected numbers of cycles beginning with $N = 2 \times 10^3$ up to $N = 5.5 \times 10^5$ for a specimen with layup ID2 and manufactured with the MRCC. The behavior is comparable for all tested laminates. The dynamic stiffness for a specific cycle number gives the mean gradient of the force-displacement-hysteresis shown in the figure. The different hystereses show that with increasing cycle numbers, the gradient and, with that, the dynamic stiffness decreases. At the same time, the specimen also exhibits creep, which is indicated by the horizontal shift of the hystereses with increasing cycle number.

Comparing the two measurement values, maximum specimen temperature and normalized dynamic stiffness, an indirect correlation is observed (Fig. 7.6b). The increasing temperature indicates damage progression inside the laminate, accompanied by friction and consequent heat generation, while the accumulation of damage leads to a decrease in dynamic specimen stiffness. Furthermore, the figure marks the number of cycles at which crack initiation and steel facesheet failure are detected. The data shows that the dynamic stiffness is a quantitative parameter that represents the state of damage of an FML specimen.

7.4 Residual Stress Influence on Fatigue Life Phases

The most significant parameter from the previous experimental test methods is the dynamic stiffness, which includes all effects over the entire specimen thickness. Therefore, Fig. 7.7 illustrates the evolution of the normalized dynamic stiffness (normalized on the initial values) over the number of cycles. Additionally, the cycle numbers correlating with crack initiation and facesheet tear indicated by the symbols are added to Fig. 7.7. For all specimens, cracks are initiated preliminary in the MRCC specimens except for layup ID4, where the crack initiation happens almost at the

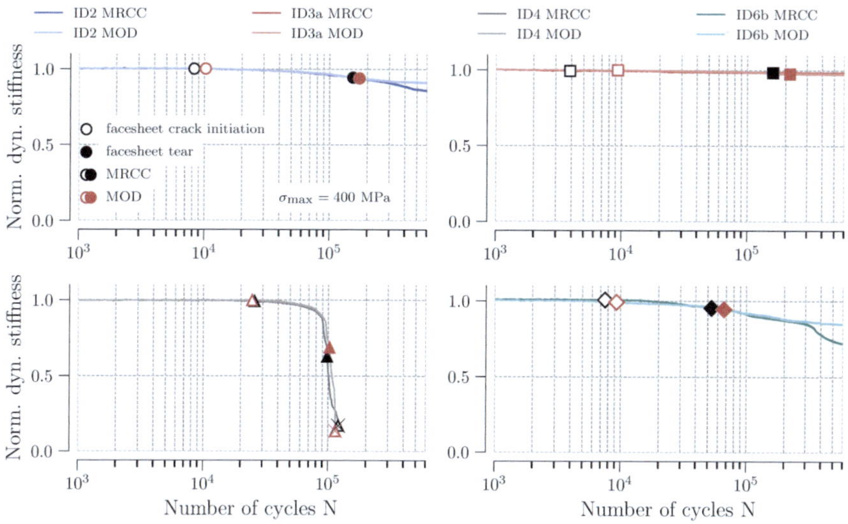

Fig. 7.7: The reduction of the residual stress state with the MOD cycle leads to an increased number of cycles until facesheet crack initiation and to an increased number of cycles until facesheet failure for all specimens except for ID4.

same cycle number for both residual stress states. The identical correlation applies to the facesheet failure. The facesheets of all MRCC specimens (except for ID4) fail before the facesheets in the MOD specimens. The plot further shows that the dynamic stiffness of each specimen starts to decrease just after cracks in the facesheets are initiated. The deviation of the MRCC and MOD curves for specimens ID2 and ID6b indicates an influence of the cure cycle after facesheet failure. To illustrate this even better, Fig. 7.8 shows the normalized dynamic stiffness with a linear x-axis to emphasize the higher cycle numbers, where the difference becomes more evident. For both layups (ID2 and ID6b), the dynamic stiffness significantly decreases after 3×10^5 cycles for the MRCC specimen, while for the MOD specimens, the dynamic stiffness seems to stabilize. From the thermography images (Fig. 7.5), it can be shown that the additional stiffness drop in the MRCC specimens is due to the failure of the inner metal sheets.

For specimens with layup ID4, no influence of residual stress on the evolution of fatigue damage is observed. This is expected since the absolute residual stress level in the laminates is small and, hence, also the relative difference between MRCC and MOD specimens (Table 5.8). For specimens with layup ID3a, no considerable reduction in dynamic stiffness is observed after facesheet failure. This can be attributed to the high laminate thickness and the small contribution of the facesheets to the overall laminate stiffness.

The following two subsections provide a detailed analysis of how residual stress states influence the two phases of fatigue life: crack initiation and crack growth.

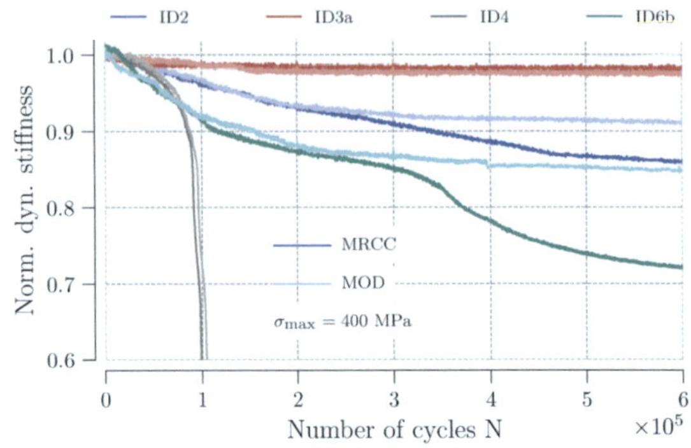

Fig. 7.8 The evolution of the dynamic stiffness over the number of cycles shows significant differences for identical layups with differing residual stress states, especially after facesheet failure for specimens ID2 and ID6b (adapted from [32]).

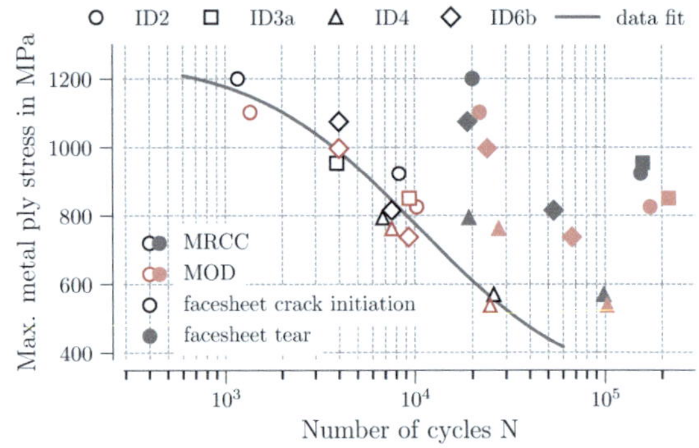

Fig. 7.9 The cycle number where crack initiation occurs correlates well with the maximum effective stress in the metal plies (adapted from [32]).

7.4.1 Crack Initiation

It was shown that crack initiation occurs earlier in the facesheets of laminates with an identical layup but higher residual stress states. The literature [27] shows that crack initiation in FML is defined by the stress concentration around the notch. Stress concentration is determined by the geometry of both the laminate and the notch, as well as the relevant stresses in the notched area. These stresses represent a combination of mechanical and thermal stresses, along with a concentration of stress influenced by the shape of the notch.

Since the specimen and notch geometry are identical for all tested specimens, only the stress dependence of the crack initiation in the FML layups is relevant. Therefore, Fig. 7.9 correlates the cycle numbers at which crack initiation is detected with the maximum effective stress in the metal ply for each laminate configuration and tested stress level. The maximum effective stress is a superposition of the thermal residual stresses and the stresses due to the maximum external load during the fatigue tests. The stresses are calculated using the analytical correlations, provided in Sec. 2.2.3. To further illustrate the dependence of the cycle number of crack initiation $N_{\text{c.i.}}$ on the maximum metal ply stress $\sigma_{\text{max, 1, steel}}$, a power function is fitted through the

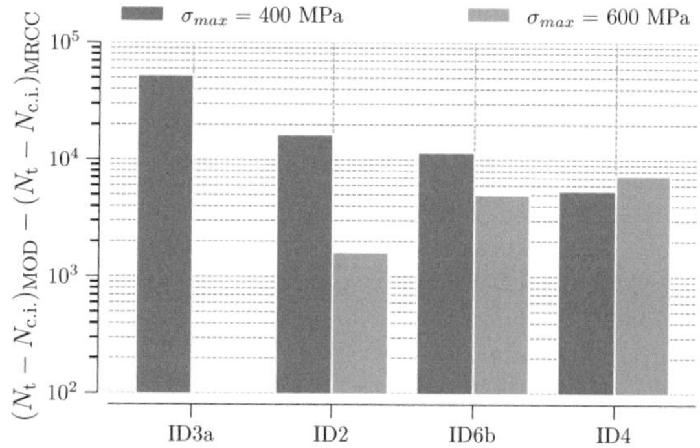

Fig. 7.10 The number of cycles between crack initiation ($N_{c.i.}$) and facesheet tear (N_t) increases for every laminate manufactured with the MOD cycle when compared to the specimens manufactured in the MRCC, indicated by positive cycle numbers in the figure (adapted from [32]).

experimental data:

$$\sigma_{\max,1,\text{steel}} = a(b + N_{\text{c.i.}})^{-1} + c \tag{7.2}$$

where a, b, and c are the fitting parameters. The data fit using this function and shown in Fig. 7.9 demonstrates a coefficient of determination of 0.9. Therefore, it can be argued that a strong correlation exists between crack initiation and the maximum effective stress in the metal plies of the laminates, which agrees with results found in fatigue investigations on GLARE [1,27].

7.4.2 Crack Growth

From Fig. 7.9, no correlation between the failure of the steel facesheets and the maximum effective metal ply stress is obvious. However, it can be shown that the number of cycles between crack initiation and facesheet failure increases for all specimens and loading conditions when reducing the residual stresses. Figure 7.10 illustrates this finding by evaluating the difference in the number of cycles between crack initiation and facesheet tear for the MRCC and MOD specimens. The specimens are sorted in order of absolute residual stress levels, with specimens ID3a possessing the highest residual stresses and specimen ID4 the lowest. For the tests with a maximum stress of 400 MPa, the influence of the residual stresses gets smaller with decreasing residual stress levels in the metal plies. The tests with a maximum stress of 600 MPa show a reverse relation. However, with the number of data points available, interpretation is difficult.

In contrast to crack initiation, the crack propagation rate can be attributed to the normalized maximum stress, a parameter also used in [17]. In Fig. 7.11, the crack propagation rate is expressed by the number of cycles between crack initiation and facesheet tear ($N_t - N_{c.i.}$). On the ordinate, the maximum effective metal ply stress $\sigma_{\max,1,\text{steel}}$ is normalized to the ultimate laminate strength $\sigma_{\text{ult,lam}}$ for each layup. The experimentally determined ultimate laminate strength from quasi-static tensile tests can be found in Fig. 3.9 for the laminates ID2, ID4, and ID6b. Layup ID3a could not

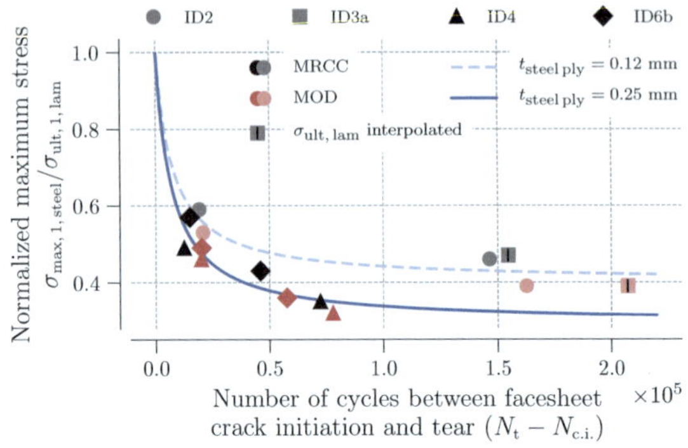

Fig. 7.11 The crack propagation in the laminates is dependent on the normalized maximum metal ply stress while differences exist for layups with differing metal sheet thicknesses (adapted from [32]).

be tested until the final laminate failure due to load restrictions of the tensile testing machine. Therefore, the values for layup ID3a result from a linear interpolation of the stress-strain data between 0.75% and an assumed ultimate failure strain of 1.75%. The ultimate failure strain is comparable for all other tested layup configurations and rather determined by the fiber material than the layup itself [1] and, therefore, reasonable to be used for interpolation. Consequently, the ultimate strength of layup ID3a is (2046 ± 71) MPa for the MRCC specimens and (2162 ± 93) MPa for the MOD specimens.

Again, a power function with fitting parameters a, b, and c of the same type as previously is applied:

$$\frac{\sigma_{\text{max, 1, steel}}}{\sigma_{\text{ult, 1, lam}}} = \frac{a}{b + (N_t - N_{\text{c.i.}})} + c \tag{7.3}$$

Figure 7.11 shows that a difference between the laminates with different metal sheet thicknesses exists. For the laminates with thinner steel sheets (layups ID2 and ID3a), it takes longer (higher number of cycles) for the crack to propagate from the edge of the notch to the edge of the specimen for a similar magnitude of the same normalized maximum metal ply stress. One reason for this effect can potentially be found in the fiber-bridging mechanism, which is more effective in laminates with thinner metal sheets. The fiber bridging effect may also be responsible for the fact that cracks propagate first in the outer metal layers, and the inner metal layers are only affected by cracks later. While the inner layers of the layups are supported by fiber layers from both sides, the facesheets are only supported by bridging fibers on one side. Similar influences of the metal sheet thickness on crack growth rate were found by [33] in GLARE.

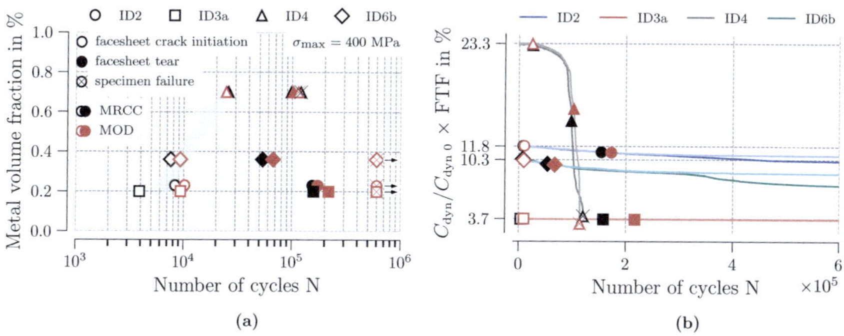

Fig. 7.12: (a) The number of cycles until facesheet crack initiation seems to correlate with the metal volume fraction in the specimens. **(b)** The dynamic stiffness decreases much faster after the failure of the facesheets with increasing facesheet thickness fraction (adapted from [32]).

7.5 Laminate Parameters Influencing Fatigue Damage Evolution

Besides the residual stress state in the laminates, three additional parameters that depend on the laminate layup are investigated for their contribution to fatigue life. These parameters are the metal volume fraction (MVF), the facesheet thickness fraction (FTF), and the number of interfaces (NoI). While the latter did not show any correlation with the fatigue damage behavior, the two other parameters are discussed in detail in the following.

Metal Volume Fraction

The fatigue damage events and their corresponding cycle count for each layup and residual stress state against the corresponding metal volume fraction are shown in Fig. 7.12a. It can be seen that the crack initiation in the facesheets occurs before 1×10^4 cycles for all layups except for ID4. The results suggest a correlation between crack initiation and the metal volume fraction, which is indicated by the arrow in the figure. However, since the residual stresses in the metal layers decrease with a higher metal volume fraction, the two effects cannot be investigated separately.

Facesheet Thickness Fraction

In Fig. 7.12b, the normalized dynamic stiffness and facesheet thickness fraction (FTF) are correlated with the number of cycles at which significant damage events occur. The results show that for high FTFs (ID4), the crack initiation in the facesheets results in a substantial decrease in the dynamic stiffness of the laminate. For lower FTFs, the influence is less pronounced. For specimens ID2 and ID6b, the dynamic stiffness continuously decreases with increasing damage accumulation in the facesheets. An additional drop in stiffness is observed after the facesheets tear,

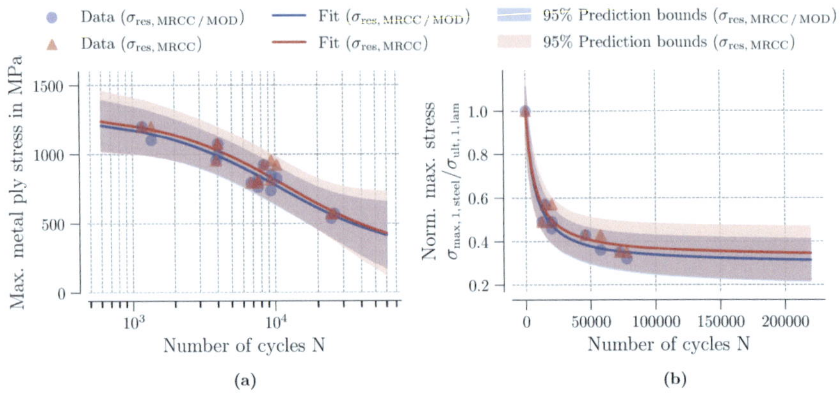

Fig. 7.13: The consideration of the individual residual stress states ($\sigma_{\text{res, MRCC/MOD}}$) when calculating the effective metal sheet stresses reduces the width of the 95%-prediction bounds for the respective stress-cycle curves when compared to data where the individual stress state is not considered ($\sigma_{\text{res, MRCC}}$) and identical residual stresses (MRCC) are assumed for identical layups regardless of the cure cycle for: **(a)** the crack initiation phase and **(b)** the crack propagation phase.

which can be attributed to the failure of the inner metal sheets based on the thermographic images. Meanwhile, for low FTFs (ID3a), no significant change in dynamic stiffness is observed even after the facesheets tear. Since the facesheets fail first in the investigated laminates, the proportion of the facesheets in the overall composite is decisive for the extent of the dynamic stiffness reduction after cracking and failure of the facesheets.

7.6 Conclusion and Consequences for SHM

The experimental fatigue tests show the evolution of the fatigue damage of notched CFRP-steel laminates under constant amplitude tension-tension loading. DIC data reveals that cracks initiate at the positions of stress concentrations at the edge of the notch in the steel facesheets. The cracks grow with the simultaneous formation of delaminations until the outer layers fail. The inner metal layers only fail at higher cycle numbers, as shown by thermographic images. With increasing damage, the specimen loses its dynamic stiffness and its temperature increases.

When comparing the cycle numbers of identical specimens but with different residual stress states, it is found that reduced residual stresses lead to increased cycles until crack initiation and also reduce the crack growth rate. Therefore, a considerable contribution of modified cure to fatigue life in CFRP-steel laminates is assumed. To further illustrate this, the correlation of the previously determined stress-dependent parameters with fatigue crack initiation and crack propagation are shown in Fig. 7.13. The figure further provides a data fit using the previously introduced

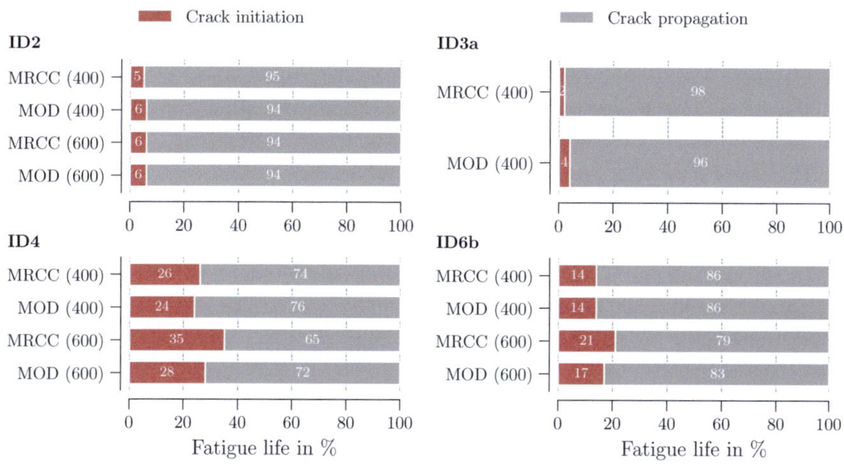

Fig. 7.14: The comparison of the contributions of crack initiation and crack propagation to the fatigue life of CFRP-steel laminates with open holes in constant amplitude fatigue tests shows that crack propagation has the largest contribution to fatigue life.

power functions and an estimation of the prediction bounds for the given data and fit. To illustrate the influence of the residual stress state on the data, the fitting functions and prediction bounds are calculated in two ways. The first calculation method is identical to the results described above, in which the residual stress state from the previous chapters was included in the calculation. The second calculation method neglects the residual stress state in the specimens. It assumes an identical residual stress state (residual stress state of MRCC specimens) for identical laminate layups independent of the cure cycle. The calculated 95%-prediction bounds show that the width of the prediction bounds for the first calculation method, i.e. taking into account the actual residual stress states, are smaller. This indicates that the parameter estimate is more accurate when the residual stress state is considered. It can, therefore, be concluded that the residual stress state plays a significant role in both crack initiation and crack propagation. Furthermore, when using the bounds as a design criterion or for residual life predictions, although the lower bound is identical for both predictions when the residual stress state of a specimen is known, an increased cycle number is predicted for a reduced residual stress state.

With these results, the fourth hypothesis, *the reduction of cure cycle-induced residual stresses improves the fatigue life of CFRP-steel laminates under constant amplitude tension-tension fatigue loading*, can be confirmed.

The results further show that fatigue life depends not only on the maximum metal ply stress level but also on the laminate architecture. In particular, the thickness of the metal sheets and the proportion of the outer metal layers in the total thickness of the laminate (facesheet thickness fraction) also determines the fatigue life.

Looking at the entire fatigue life of the CFRP-steel laminates investigated, the results prove that, in contrast to monolithic metals, the crack growth phase under cyclic loading takes up most of the fatigue life. This is illustrated in Fig. 7.14 for all investigated layups and the different residual and maximum stress levels tested. For layup ID 4 with high residual stresses and a maximum stress level of 600 MPa, the crack initiation phase accounts for 35% of the fatigue life. For all other layups and configurations, the crack initiation phase takes up significantly less than a third of the fatigue life. This finding agrees well with investigations in GLARE [9] and demonstrates the fatigue-tolerant properties of the investigated CFRP-steel laminates.

This characteristic of CFRP-steel laminates and FML, in general, shows that a GUW-based SHM system is particularly suitable for this type of material, as fatigue damage (micro-cracks) does not have to be detected as soon as it occurs. Due to the slow crack growth rate, it is sufficient if cracks are detected from a significant length (macro-cracks). This simplifies the design of the GUW-based SHM system since macro-cracks are assumed to have a higher interaction with a passing wavefront due to their size compatibility. Furthermore, the laminates examined in this study (except ID4) have shown that they still possess a high residual stiffness and, thus, residual service life even after significant damage (facesheet tear). The dynamic stiffness after the failure of the facesheets is still above 90% for laminates ID2, 3a, and 6b.

Regarding service life prediction for CBM approaches, using the residual stress state after manufacturing as an input for a predictive algorithm likely increases its accuracy. This finding is independent of the method used for fatigue life prediction. This also means that knowledge of the residual stress state is crucial for accurately predicting the remaining service life within a robust SHM system after damage detection.

In future work, the results must be validated for additional load levels, including tension/compression, under variable amplitude tests and for more laminate layups. In particular, FMLs with facesheets made of the fiber constituent and laminates with more than a single fiber direction are of interest. Furthermore, in this work, the focus was on fatigue crack propagation, whereas the damage observed consists of simultaneous fatigue cracks and delaminations, which need further attention in future studies. Finally, future cyclic tests should focus on the final laminate failure to investigate the influence of the residual stresses and other parameters on the FML residual strength.

References

1. Alderliesten, R.: Fatigue and fracture of fibre metal laminates, *Solid mechanics and its applications*, vol. 236. Springer, Cham (2017)
2. Alderliesten, R.C.: Fatigue crack propagation and delamination growth in glare. Dissertation, Technische Universiteit Delft, Netherlands (2005)
3. Alderliesten, R.C.: On crack tunneling and plane-strain delamination in laminates. International Journal of Fracture **148**(4), 401–414 (2007). doi:10.1007/s10704-008-9212-8
4. Both, J.C.: Tragfähigkeit von CFK-Metall-Laminaten unter mechanischer und thermischer Belastung: Zugl.: München, Techn. Univ., Diss., 2014. Ingenieurwissenschaften. Dr. Hut, München (2014). URL `http://mediatum.ub.tum.de/node?id=1173300`

5. Burgers, A., Kempen, P.D.: Automatic crack length measurements by the electrical potential drop method with computer control: Report LR-309. URL `http://resolver.tudelft.nl/uuid:92d25f8b-960a-4192-865a-a7ea3ad0d377`
6. Federal Aviation Administration: Damage Tolerance and Fatigue Evaluation of Structure
7. Foote, P.: New Guidelines for Implementation of Structural Health Monitoring in Aerospace Applications. SAE International Journal of Aerospace **6**(2), 525–533 (2013). doi:10.4271/2013-01-2219
8. Hausmann, J., Schmidt, S.: Thermal residual stresses in thermoplastic CFRP-steel laminates: Modification and influence on fatigue life. Journal of Micromechanics and Molecular Physics **06**(03), 49–55 (2021). doi:10.1142/S2424913021430050
9. Homan, J.J.: Fatigue initiation in fibre metal laminates. International Journal of Fatigue **28**(4), 366–374 (2006). doi:10.1016/j.ijfatigue.2005.07.030
10. Khan, S.U., Alderliesten, R.C., Benedictus, R.: Post-stretching induced stress redistribution in Fibre Metal Laminates for increased fatigue crack growth resistance. Composites Science and Technology **69**(3-4), 396–405 (2009). doi:10.1016/j.compscitech.2008.11.006
11. Kim, H.S., Park, S.W., Hwang, H.Y., Lee, D.G.: Effect of the smart cure cycle on the performance of the co-cured aluminum/composite hybrid shaft. Composite Structures **75**(1-4), 276–288 (2006). doi:10.1016/j.compstruct.2006.04.030
12. Kim, H.S., Yoo, S.H., Chang, S.H.: In situ monitoring of the strain evolution and curing reaction of composite laminates to reduce the thermal residual stress using FBG sensor and dielectrometry. Composites Part B: Engineering **44**(1), 446–452 (2013). doi:10.1016/j.compositesb.2012.04.021
13. Koord, J., Stüven, J.L., Petersen, E., Völkerink, O., Hühne, C.: Investigation of exact analytical solutions for circular notched composite laminates under tensile loading. Composite Structures **243**, 112180 (2020). doi:10.1016/j.compstruct.2020.112180
14. Koord, J., Stüven, J.L., Völkerink, O., Petersen, E., Hühne, C.: Investigation of exact analytical solutions for composite laminates under pin-bearing loading. Composite Structures **292**, 115605 (2022). doi:10.1016/j.compstruct.2022.115605
15. Laliberté, J., Mahendran, M., Djokic, D., Li, C., Kratz, J.: Effect of process-induced residual stresses on mechanical properties and fatigue crack initiation in fibre metal laminates. In: L. Lazzeri (ed.) ICAF 2007 : durability and damage tolerance of aircraft structures: metals vs composites; proceedings of the 24th symposium of the International Committee on Aeronautical Fatigue, pp. 576–592. Pacini Ed., Pisa (2007)
16. Lekhnitskii, S.G.: Anisotropic plates: translated from the second Russian edition by S.W. Tsai and T. Cheron. Gordon and Breach, New York (1968)
17. Mandell, J.F., Samborsky, D.D., Wang, L., Wahl, N.K.: New Fatigue Data for Wind Turbine Blade Materials. In: ASME 2003 Wind Energy Symposium, pp. 167–179. ASMEDC (2003). doi:10.1115/WIND2003-692
18. Mrzljak, S., Schmidt, S., Kohl, A., Hülsbusch, D., Hausmann, J., Walther, F.: Testing procedure for fatigue characterization of steel-CFRP hybrid laminate considering material dependent self-heating. Materials (Basel, Switzerland) **14**(12) (2021). doi:10.3390/ma14123394
19. Mrzljak, S., Trautmann, M., Blickling, P., Wagner, G., Walther, F.: Fatigue condition monitoring of notched thermoplastic-based hybrid fiber metal laminates using electrical resistance measurement and digital image correlation. Journal of Composite Materials **57**(17), 2669–2687 (2023). doi:10.1177/00219983231176257
20. Mrzljak, S., Trautmann, M., Wagner, G., Walther, F.: Influence of aluminum surface treatment on tensile and fatigue behavior of thermoplastic-based hybrid laminates. Materials (Basel, Switzerland) **13**(14) (2020). doi:10.3390/ma13143080
21. Mrzljak, S., Trautmann, M., Wagner, G., Walther, F.: Very high cycle fatigue assessment of thermoplastic-based hybrid fiber metal laminate by using a high-frequency resonant testing system. International Journal of Fatigue p. 108361 (2024). doi:10.1016/j.ijfatigue.2024.108361
22. Rauter, N.: Analyse des Einflusses der Werkstoffdegradation auf die nichtlineare Wellenausbreitung in unidirektionalen Compositen: Experimentelle Untersuchung mit numerischer Analyse und Modellbildung. Ph.D. thesis, Universitätsbibliothek der HSU / UniBwH (2017). doi:10.24405/4183

23. Reifsnider, K.L.: Damage and Damage Mechanics. In: Fatigue of Composite Materials, *Composite Materials Series*, vol. 4, pp. 11–77. Elsevier (1991). doi:10.1016/B978-0-444-70507-5.50006-8
24. Savin, G.N.: Stress distribution around holes. Nasa Technical Translation (1968)
25. Shim, D.J., Alderliesten, R.C., Spearing, S.M., Burianek, D.A.: Fatigue crack growth prediction in GLARE hybrid laminates. Composites Science and Technology **63**(12), 1759–1767 (2003). doi:10.1016/S0266-3538(03)00082-4
26. Soutis, C., Filiou, C.: Stress distributions around holes in composite laminates subjected to biaxial loading. Applied Composite Materials **5**(6), 365–378 (1998). doi:10.1023/a:1008837715030
27. Spronk, S., Şen, I., Alderliesten, R.C.: Predicting fatigue crack initiation in fibre metal laminates based on metal fatigue test data. International Journal of Fatigue **70**, 428–439 (2015). doi:10.1016/j.ijfatigue.2014.07.004
28. Takamatsu, T., Matsumura, T., Ogura, N., Shimokawa, T., Kakuta, Y.: Fatigue crack growth properties of a GLARE3-5/4 fiber/metal laminate. Engineering Fracture Mechanics **63**(3), 253–272 (1999). doi:10.1016/S0013-7944(99)00021-1
29. Trautmann, M., Mrzljak, S., Walther, F., Wagner, G.: Mechanical properties of thermoplastic-based hybrid laminates with regard to layer structure and metal volume content. Metals **10**(11), 1430 (2020). doi:10.3390/met10111430
30. Vlot, A., Gunnink, J.W.: Fibre metal laminates: An introduction. Springer Netherlands, Dordrecht (2001). URL `https://ebookcentral.proquest.com/lib/gbv/detail.action?docID=3565888`
31. Wiedemann, J., Mrzljak, S., Hühne, C., Walther, F.: Increased accuracy of service life prediction for fiber metal laminates by consideration of the manufacturing-induced residual stress state: Proceedings of the 10th European Workshop on Structural Health Monitoring (EWSHM 2024). e-Journal of Nondestructive Testing **29**(7) (2024). doi:10.58286/29832
32. Wiedemann, J., Mrzljak, S., Koord, J., Hühne, C., Walther, F.: Modified cure cycles for increased fatigue performance of fiber metal laminates. Composite Structures **351**, 118631 (2025). doi:10.1016/j.compstruct.2024.118631
33. Wilson, G.S.: Fatigue crack growth prediction for generalized fiber metal laminates and hybrid materials. [s.n.], [S.l.] (2013). URL `http://resolver.tudelft.nl/uuid:8925fc2a-47e8-457a-960c-f86b5065fa60`

Chapter 8
Conclusion and Outlook

8.1 Main Contributions of the Dissertation

Operating a structural health monitoring (SHM) system within fiber metal laminates (FMLs) presents distinct challenges unique to this type of material. Combining fiber-reinforced polymers (FRPs) with metals in FMLs induces significant residual stresses after manufacturing. Specifically, in unidirectionally oriented FMLs, the residual stress state is complex, characterized by tensile stresses in the metal layers and compressive stresses in the fiber layers. A central concern for a guided ultrasonic wave (GUW) based SHM system is whether these residual stress states impact its functionality.

To address this question, the residual stress states of various FML layups and for different cure cycles were comprehensively characterized and quantified using sensors integrated into the manufacturing process and asymmetric specimens. The quantified residual stress states served as the foundation for investigating their influence on the SHM system. The work primarily focused on the interaction between residual stresses and the propagation of fundamental GUW modes, as well as the effect on the fatigue life of FMLs.

This research concentrated on FMLs composed of carbon fiber-reinforced polymer (CFRP) and steel layers, a combination extensively documented in the literature due to its wide range of applications. Additionally, the combination of CFRP with steel results in significantly higher residual stress levels than for other FML configurations due to the substantial differences in the coefficients of thermal expansion (CTE) and the high stiffness of the materials involved. Consequently, this type of FML is particularly suitable for investigating the effects of residual stress states on an SHM system. Nonetheless, the findings from this study are expected to apply to FMLs incorporating other FRPs and metals.

The main contributions of this dissertation are discussed in detail in the following paragraphs.

J. Wiedemann, *Relevance of Manufacturing-Induced Residual Stress States in Fiber Metal Laminates for Structural Health Monitoring*, Mechanics and Adaptronics, https://doi.org/10.1007/978-3-032-09925-9_8

Residual Stress Quantification

This work developed a method that uses strain gauges to record the strains arising during cure throughout the entire manufacturing process [4,9]. This allows all strains between the start and the end of the process to be quantified. The residual strains at the end of manufacturing can be converted directly into residual stress using CLT (Eq. (4.3)). If the experimental hardware used in future tests is identical to the one used in this work, reference measurements using a dummy gauge can be skipped. Instead, a master compensation curve (Eq.(4.2) and Fig. 4.3) to compensate for the temperature influence is available. The comparison with FBG strain data shows that such a method is unique for FML and cannot be used for FRPs. For FRPs, a reference temperature like bonding or stress-free temperature is necessary to quantify residual stresses.

The experimental strain measurements demonstrated a significant difference between the bonding (T_{bonding}) and stress-free (T_{sf}) temperatures, often treated the same in the existing literature. However, this work demonstrates that the distinction between T_{bonding} and T_{sf} is crucial: strain gauge measurements suggest that if T_{bonding} is used as a reference temperature, the calculation of residual stresses becomes less accurate, especially at low bonding temperatures. The reason is that non-linearities in the expansion behavior for low bonding temperatures are not considered when $T_{\text{sf}} = T_{\text{bonding}}$ is used. In contrast, these non-linearities are automatically considered in other methods that directly determine T_{sf}, e.g., using asymmetric specimens.

Besides residual stress quantification with strain gauges, another method that is better applicable to industrial manufacturing is presented [10]. This method requires only simple test specimens, which can be manufactured simultaneously with a relevant component. After manufacturing, different values can be determined from the curvature of these test specimens, independent of the component. Recommendations for the design of the test specimens are given, and the influence of intrinsic and extrinsic parameters on the curvature is quantified. The results show that the curvature is influenced by the layer structure (intrinsic), the choice of tool material (extrinsic), and the cure cycle (extrinsic). At the same time, T_{sf} is only influenced by extrinsic parameters (tool material and cure cycle). This finding allows the transferability of T_{sf} to other laminates manufactured within the same process. To apply this method in future works, the novel approach is presented (illustrated in Fig. 5.17) with relevant equations.

Residual Stress Influence on Guided Ultrasonic Wave Propagation

A method originally employed for isotropic materials to measure dispersion diagrams for the fundamental GUW modes A_0 and S_0 over a large frequency range was validated for its use in FMLs [1]. For the first time, dispersion diagrams for the two fundamental GUW modes were experimentally determined for different FML layup configurations up to a frequency-thickness product of 2 MHz mm. Repeated measurements on identical and different FML specimens proved the error between different measurements to be below 2%. Comparing the data with analytical solutions suggests that the analytical relations known from homogeneous materials also apply to GUW propagation in FMLs [2].

Measurements on laminates with differing manufacturing-induced residual stress states indicate that the influence of the residual stress on the phase velocity of the fundamental GUW modes is small within the investigated laminates [5]. The comparison of FMLs with especially high residual stress levels demonstrates that the potential effect is smaller than the uncertainties resulting from using two different test specimens. These uncertainties can be attributed to uncertainties in the material itself and to uncertainties due to the readjustment of the experiment after changing a specimen. The results suggest that residual stresses are of minor importance for SHM in the investigated laminates, the used frequencies, and the two fundamental GUW modes.

In contrast, additional experiments with prestressed FML specimens revealed that prestress significantly affects the phase velocities of the fundamental A_0- and S_0-mode. For both modes, a linear correlation between an increase in prestress and an increase in velocity is observed in the investigated frequency range. This is particularly interesting as non-linear elastic theory predicts an inversely proportional relationship between prestress and phase velocity change when using Murnaghan or Neo-Hooke material models for low frequencies in isotropic materials. Furthermore, for the A_0-mode, the relative change in phase velocity is greater than 5% for frequency-thickness products below 100 kHz mm, which is a relevant frequency for SHM applications. Consequently, unlike residual stress, the influence of prestress in FML seems more relevant for SHM.

Residual Stress Influence on Fatigue Life

The experimental results show that GUW-based SHM systems are suitable for FMLs due to the material's fatigue characteristics. Crack growth in the investigated CFRP-steel FML is a slow process that takes up most of the fatigue life. In addition, the results have shown that the failure of individual steel layers is not necessarily accompanied by component failure, and considerable residual strength of the laminate is maintained [6]. Therefore, detecting macro-damage seems sufficient to monitor the structural integrity of a component instead of monitoring the damage initiation phase.

Different residual stress states introduced by the manufacturing process in the tested laminates influence their fatigue life [7]. Reduced residual stresses delay crack initiation and slow down crack propagation. If the corresponding cycle numbers for crack initiation and crack propagation are correlated with the stresses present in the metal layers, an improved quality of prediction is shown when the residual stress state is considered. This indicates that prediction models for remaining service life can be more accurate, and uncertainties due to the scattering of the results can be reduced.

The experimental results further showed that the fatigue behavior of CFRP-steel laminates is also highly dependent on the laminate layup. Characteristic variables such as the facesheet thickness fraction and the metal volume fraction also govern fatigue life. Furthermore, the tests demonstrated that the investigated laminates do not fail due to simultaneous damage in all metal layers. Typically, the facesheets fail first by a single crack on both sides of the notch. In contrast, the inner layers fail partly due to the propagation of multiple cracks on both sides of the notch and well after the failure of the facesheets.

8.2 Evaluation of the Research Hypotheses

Based on the motivation presented in Chap. 1 and the state of the research discussed in Chap. 2, four research hypotheses were formulated. While Chap. 3 introduced all the relevant materials and manufacturing processes, the research hypotheses structured the investigations in the Chaps. 4-7. The focus in Chaps. 4 and 5 was on characterizing the residual stress states in CFRP-steel laminates, which is the basis for the subsequent assessment of their influence on the GUW propagation in Chap. 6 and fatigue characteristics in Chap. 7. The formulated research hypotheses can be evaluated based on the individual results of the respective chapters. Table 8.1 shows whether the individual hypotheses are accepted, partially accepted, or rejected.

While the first and fourth hypotheses are accepted, the second is only partially accepted. The results clearly show that the stress-free temperature of such asymmetric specimens is only sensitive to extrinsic process parameters, in contrast to the curvature, which is also influenced by the layup. Therefore, transferring the stress-free temperature to any other layup is logical. However, to fully accept the hypothesis, additional studies are necessary that unconditionally confirm this correlation for various symmetric layup configurations using both methods for residual stress quantification presented in this work.

Table 8.1: Evaluation of the research hypotheses formulated in this work. The colors indicate whether the hypothesis is accepted (green), partially accepted (yellow), or rejected (red).

	Research hypothesis	Evaluation
1.	*In contrast to FRPs, strain sensors attached to the metal layers of FMLs enable the recording of meaningful time-based strain data from the beginning of the manufacturing process.*	● ○ ○ (green)
2.	*Asymmetric specimens can determine the residual stress states in FMLs with any dimension and layer structure by deriving a parameter only sensitive to extrinsic manufacturing process variables.*	○ ● ○ (yellow)
3.	*Cure cycle-induced residual stress states influence the phase velocity of the fundamental GUW modes (A_0 and S_0) propagating in an FML.*	○ ○ ● (red)
4.	*The reduction of cure cycle-induced residual stresses improves the fatigue life of CFRP-steel laminates under constant amplitude tension-tension fatigue loading.*	● ○ ○ (green)

The third hypothesis is rejected since the measurement setup could not detect a considerable influence of differing residual stress states. This does not imply that there is no effect; instead, the effect observed in the tested laminates is less influential than the uncertainties introduced by using different test specimens. Therefore, the residual stress influence on an SHM system is considered of minor importance. On the other hand, additional investigations have shown that an externally applied prestress on FMLs considerably influences the phase velocities of propagating GUWs in relevant frequency ranges and, therefore, requires special attention in future studies.

8.3 Open Questions and Future Work

This dissertation has addressed and clarified open research questions in the context of an SHM system for FMLs. However, it has also raised new questions that are beyond the scope of this dissertation and require further research in future studies. Some of these new questions emerge from combining and expanding the three previously introduced categories. Others are independent of the categories and arise from applying findings from the coupon level to the component level:

- *Residual stress in complex laminates*: Future investigations should focus on laminates from practical use cases. It is unclear how residual stresses are distributed in laminates with varying MVF, ply discontinuities, splices, or varying thicknesses. Furthermore, the transferability of the stress-free temperature identified from asymmetric samples to these considerably more complex and symmetric laminates must be further validated, as previously discussed.
- *Damage mechanisms and fatigue behavior in complex laminates*: The damage mechanisms and fatigue damage associated with more complex laminates require further investigation. In particular, there has been limited research on the behavior of transition zones and bolted connections under fatigue loads. In particular, the fatigue behavior of an FML when the load is applied via a bolt is of great interest. Furthermore, additional damage scenarios relevant to the application and material combination need further investigations (e.g., low- and high-velocity impact, environmental degradation, hole elongation, and delamination).
- *Conflicting objectives in laminate production*: With regard to laminate production, there is a conflict of objectives between the requirements of industrial production and the performance of the components. While the process requirements aim for short, cost-effective, simple, and reproducible processes, they contradict the longer modified cure cycles, which enable higher strength and improved fatigue life. Furthermore, the extent to which modified cure cycles can be used in real production environments remains questionable. Challenges include large autoclaves, inaccurate temperature control, and significant temperature deviations within industrial autoclaves, as discussed in [8].
- *Advanced methods for residual stress quantification*: Other methods for residual stress quantification are worth investigating. X-ray diffraction (XRD) measurements are promising for local residual stress quantification in a laminate when no

information about the manufacturing process is available. This method allows for the non-destructive measurement of the residual stress state at specific locations post-manufacturing. It is especially relevant for large components where local process temperatures cannot be recorded. The potential of the XRD technique has already been highlighted by the author [7]. Additionally, new distributed fiber optic measurement systems enable continuous strain monitoring along the entire fiber length, demonstrating great potential for assessing the homogeneity of residual stresses in one or more dimensions.

- *GUW propagation in complex laminates*: SHM using GUWs in laminates with increased inhomogeneity poses significant challenges. It is currently unclear how features such as ply discontinuities and areas with varying MVFs interact with propagating GUWs. Stiffness discontinuities may act as reflectors for a passing wavefront, complicating damage detection. Given these challenges, the monitoring strategy used in the SHM system becomes critically important. Factors such as the number and position of actuators and sensors and the type of excitation require careful consideration to ensure effective monitoring of these FMLs. To validate damage detection using SHM, the combination with other non-destructive monitoring techniques like photo-thermal defect imaging [3] seems promising.
- *Parameter uncertainties for modeling GUWs*: The influence of material uncertainties and other variations caused by the manufacturing process must be comprehensively investigated. The material parameters play a decisive role in the model-based description of the wave propagation. When using hyperelastic material models, additional elasticity parameters are required, which cannot be readily determined. It is, therefore, essential to understand the interactions between parameter uncertainty and wave propagation to correctly identify the appropriate characteristic values needed in those models.
- *Temperature and residual stress*: The extension of the investigations into the influence of residual stresses by considering the influence of temperature is very promising. A temperature change not only addresses the temperature-related sensitivity of wave propagation but also changes the residual stress state. If it is possible to separate the two effects, more significant residual stress differences may be achievable than those generated by the manufacturing process in this work. Furthermore, it remains open whether a potential impact of the residual stresses can be described by the theoretical correlations.
- *Prestress influence on GUW propagation*: The additional investigations into the influence of prestress on GUW propagation have revealed that this effect is considerably more pronounced in the evaluated FMLs than what is typically observed in isotropic materials. This influence impacts the A_0- and S_0-mode within frequency ranges relevant for SHM. Consequently, this effect cannot be overlooked when operating SHM systems for FMLs. It necessitates further research, including analytical and numerical modeling, to fully understand and describe its implications.

This dissertation has shown why an SHM system is promising for monitoring the structural integrity of an FML during operation, taking into account its complex damage characteristics. The results must now be transferred from the coupon to the component level. In particular, the combination of GUW measurements and damage analysis is important to advance the technological maturity of an SHM system. This work significantly contributed by demonstrating fundamental relationships between residual stress states and relevant variables of an SHM system and by quantifying relevant effects.

References

1. Barth, T., Wiedemann, J., Roloff, T., Behrens, T., Rauter, N., Hühne, C., Sinapius, M., Lammering, R.: Experimental determination of dispersion diagrams over large frequency ranges for guided ultrasonic waves in fiber metal laminates. Smart Materials and Structures **32**(8), 085011 (2023). doi:10.1088/1361-665X/ace0ea
2. Barth, T., Wiedemann, J., Roloff, T., Hühne, C., Sinapius, M., Rauter, N.: Investigations on guided ultrasonic wave dispersion behavior in fiber metal laminates using finite element eigenvalue analysis. PAMM **23**(1) (2023). doi:10.1002/pamm.202200149
3. Gahleitner, L., Thummerer, G., Plank, B., Wiedemann, J., Mayr, G., Hühne, C., Burgholzer, P., Cakmak, U.: Photothermal defect imaging in hybrid fiber metal laminates using the virtual wave concept. Journal of Applied Physics **135**(7) (2024). doi:10.1063/5.0188449
4. Kappel, E., Prussak, R., Wiedemann, J.: On a simultaneous use of fiber-Bragg-gratings and strain-gages to determine the stress-free temperature Tsf during GLARE manufacturing. Composite Structures **227** (2019). doi:10.1016/j.compstruct.2019.111279
5. Wiedemann, J., Barth, T., Roloff, T., Kluge, T., Rauter, N., Hühne, C.: Influence of residual stresses on the dispersion behavior of guided ultrasonic waves in fiber metal laminates. In: Proceedings of the 14th International Workshop on Structural Health Monitoring. Destech Publications, Inc (2023). doi:10.12783/shm2023/36729
6. Wiedemann, J., Mrzljak, S., Hühne, C., Walther, F.: Increased accuracy of service life prediction for fiber metal laminates by consideration of the manufacturing-induced residual stress state: Proceedings of the 10th European Workshop on Structural Health Monitoring (EWSHM 2024). e-Journal of Nondestructive Testing **29**(7) (2024). doi:10.58286/29832
7. Wiedemann, J., Mrzljak, S., Koord, J., Hühne, C., Walther, F.: Modified cure cycles for increased fatigue performance of fiber metal laminates. Composite Structures **351**, 118631 (2025). doi:10.1016/j.compstruct.2024.118631
8. Wiedemann, J., Pittsyk, T., Völkerink, O., Hühne, C.: Temperature distribution inside composite and fiber metal laminates during modified cure cycles. Procedia CIRP **131**, 107–112 (2025). doi:10.1016/j.procir.2024.10.333
9. Wiedemann, J., Prussak, R., Kappel, E., Hühne, C.: In-situ quantification of manufacturing-induced strains in fiber metal laminates with strain gages. Composite Structures **Vol. 691**(1), 115967 (2022). doi:10.1016/j.compstruct.2022.115967
10. Wiedemann, J., Schmidt, J.U.R., Hühne, C.: Applicability of asymmetric specimens for residual stress evaluation in fiber metal laminates. Journal of Composites Science **6**(11), 329 (2022). doi:10.3390/jcs6110329

MIX
Papier aus verantwortungsvollen Quellen
Paper from responsible sources
FSC® C105338

If you have any concerns about our products,
you can contact us on
ProductSafety@springernature.com

In case Publisher is established outside the EU,
the EU authorized representative is:
Springer Nature Customer Service Center GmbH
Europaplatz 3, 69115 Heidelberg, Germany

Printed by Libri Plureos GmbH
in Hamburg, Germany